JAMES WATT, CHEMIST: UNDERSTANDING THE ORIGINS OF THE STEAM AGE

SCIENCE AND CULTURE IN THE NINETEENTH CENTURY

Series Editor: Bernard Lightman

TITLES IN THIS SERIES

1 Styles of Reasoning in the British Life Sciences: Shared Assumptions, 1820–1858
James Elwick

2 Recreating Newton: Newtonian Biography and the Making of Nineteenth-Century History of Science
Rebekah Higgitt

3 The Transit of Venus Enterprise in Victorian Britain
Jessica Ratcliff

4 Science and Eccentricity: Collecting, Writing and Performing Science for Early Nineteenth-Century Audiences
Victoria Carroll

5 Typhoid in Uppingham: Analysis of a Victorian Town and School in Crisis, 1875–1877
Nigel Richardson

6 Medicine and Modernism: A Biography of Sir Henry Head
L. S. Jacyna

7 Domesticating Electricity: Expertise, Uncertainty and Gender, 1880–1914
Graeme Gooday

FORTHCOMING TITLES

Natural History Societies and Civic Culture in Victorian Scotland
Diarmid A. Finnegan

Communities of Science in Nineteenth-Century Ireland
Juliana Adelman

Regionalizing Science: Placing Knowledges in Victorian England
Simon Naylor

JAMES WATT, CHEMIST: UNDERSTANDING THE ORIGINS OF THE STEAM AGE

BY

David Philip Miller

Published by the University of Pittsburgh Press, Pittsburgh, Pa., 15260

Manufactured in the United States of America
Printed on acid-free paper
10 9 8 7 6 5 4 3 2 1

Cataloging-in-Publication is available from the British Library

ISBN 13: 978-0-8229-8172-5 Hardback: 978-1-85196-974-6
ISBN 10: 0-8229-8172-6

CONTENTS

For Dorothy and Albert Miller

ACKNOWLEDGEMENTS

First I want to acknowledge the support of my colleagues in the School of History and Philosophy at the University of New South Wales. In times when universities, without wishing to deny the privileges involved in working in one, are troubled and sometimes confusing places to be, collegial encouragement makes up for a lot. My wife, Margaret, our children, Kylie and Tom, and friends have borne another Watt project with good grace, and tolerated, maybe on occasion even secretly enjoyed, my frequent and extended absences overseas in libraries and archives. During those absences I have benefited from the help of numerous librarians and archivists, especially those in: the British Library; Birmingham Central Library; the University of Glasgow, Special Collections; the Watt Monument Library in Greenock; and, perhaps rather incongruously for a project such as this, the Library of Congress in Washington D. C. The efficient services of the Interlibrary Loans Unit at the University of New South Wales Library have been invaluable, as usual.

This book has grown out of work that I have been doing over a few years now and it draws in parts on previously published articles. I thank the relevant bodies and individuals for permission to use here work previously published. Chapters 1 and 4 draw on parts of 'True Myths: James Watt's Kettle, his Condenser and his Chemistry', *History of Science*, 42 (2004), pp. 333–60, by permission of Science History Publications. Chapters 5 and 6 make use of and extend material from 'Seeing the Chemical Steam through the Historical Fog: Watt's Steam Engine as Chemistry', *Annals of Science*, 65 (2008), pp. 47–72. These extracts are reprinted by permission of the publisher (Taylor & Francis Ltd, http://www.tandf.co.uk/journals). Chapter 4 also uses material published, with Trevor Levere, in '"Inhale it and See"? The Collaboration between Thomas Beddoes and James Watt in Pneumatic Medicine', *Ambix*, 55 (2008), pp. 5–28. For this I thank my coauthor and the copyright holder, Maney Publishing (http://www.maney.co.uk/journals/ambix and http://www.ingentaconnect.com/content/maney/amb).

Alongside Trevor Levere, the following have given assistance or encouragement to me in various ways, for which I thank them: Jim Andrew, Patricia Fara, Jan Golinski, Rob Iliffe, Peter M. Jones, Christine MacLeod, Ben Marsden, Ray

McKenzie, Jack Morrell, Simon Schaffer and Larry Stewart. Thank you also to Pickering & Chatto and the Series Editor, Bernard Lightman, for allowing me to appear in this guise. Tom Miller provided invaluable technical support when he wasn't tying up my computers by running lengthy numerical simulations of granular processes.

Although my parents are, in a fundamental sense, very much to blame for all that follows, I am happy to take complete responsibility for it whilst dedicating this book to them, and to all that they stand for.

Sydney
Valentine's Day, 2009.

LIST OF FIGURES

Frontispiece: James Eckford Lauder 'James Watt and the Steam Engine: Dawn of the 19th Century' (1855). (Reproduced by Permission of the National Galleries of Scotland)

INTRODUCTION

By way of introduction I need to do two things. The first is to provide a historical outline of the life of James Watt, a general overview that will assist in situating my specific, and rather specialized, argument about the centrality of chemistry to his life and work. The second is to explain the structure and the strategy of my study, as well as some of the basic assumptions that underlie it.

The 'Great Steamer' – A Life Outlined

James Watt, who the geologist Roderick Murchison referred to in 1839, with a peculiar mixture of deference, affection and condescension, as 'the Great Steamer',[1] had been born just over a hundred years earlier, in 1736, to a moderately prosperous merchant family in Greenock, a small town on the lower reaches of the River Clyde, near Glasgow in Scotland. Watt's paternal grandfather was a 'professor' of mathematics, teaching that subject and navigation in a community dominated increasingly in the mid-eighteenth century by maritime trade. Watt's father was a ships' chandler and general merchant. While Watt's elder brother John entered the family business – he was lost at sea on a merchant venture – the family was prosperous enough to invest in an informal apprenticeship for Watt to the London-based instrument maker John Morgan. The young Watt's time in London was undoubtedly difficult. He worked exceedingly hard and suffered problems with his health. But the experience opened up for him a world of instruments, skills and trade that profoundly shaped his later career.[2]

The next step in that career was Watt's establishment as an instrument-maker at the University of Glasgow, within the confines of the College. Though Watt certainly did make and repair instruments for the professorate of the College he also engaged in a more general trade in musical instruments and fancy toys. In the early days he was also an agent for his father's business and plied the general hardware trade.[3] By all accounts, as a child Watt had been studious and thoughtful, though his education at the local grammar school had been badly interrupted by health problems. Whatever the case with his early education, after his apprenticeship he became an ardent and effective autodidact. He took full advantage during the later 1750s and early 1760s of the associations with the

Glasgow College professorate, and students, that his position at the College gave him. These were the years of his 'discipleship' with the newly appointed Professor of Chemistry, Joseph Black, from whom, in an informal fashion, Watt imbibed much of his early knowledge of chemistry, and of the chemistry of heat in particular. These were the years of Watt's first experiments with steam and with a model steam engine, famously given to him to repair by the Professor of Natural Philosophy, John Anderson in 1763. During this time too Watt befriended a number of students at the university, young men close to his own age. Among these was John Robison, his lifelong friend, and in his later years a key witness, both legal and literary, to Watt's work on the steam engine.[4]

On Glasgow Green today, in the shadow of the massive obelisk commemorating Lord Nelson, is a modest, inscribed boulder supposedly marking the spot where, in 1765, having contemplated the problems of the model Newcomen engine, Watt had the sudden inspiration of the separate condenser. His idea of condensing the steam in a separate vessel rather than in the main cylinder into which it had originally been injected led to a transformation in the efficiency of steam engines. The boulder celebrating this supposed 'Eureka!' moment was placed there only as recently as 1969, the year of the bicentenary of Watt's first steam-engine patent claiming the separate condenser as his invention. The story of the inspiration on Glasgow Green, however, goes back to the early nineteenth century. There is no direct evidence of its veracity in the Watt archives and it is one of many examples of the difficulties of separating fact and fiction in Watt's life.[5]

By the time that he took out his patent, Watt was already in a partnership with Dr John Roebuck of Kinneil, with a view to constructing a steam engine on Watt's new principles. Progress was slow and, faced with the exigencies of supporting a new family – Watt had married a cousin, Margaret Miller, in 1764, and their first child Margaret had been born in 1767, with James Jr following two years later – he abandoned the instrument and hardware business to pursue his fortune for the next few years as a journeyman civil engineer and surveyor. Watt worked on a score of projects, mainly canal and harbour surveys but also bridge and waterwheel constructions in the early 1770s.[6] It was whilst away from home, having just begun the Caledonian Canal survey, in September 1773, that Watt learned that his wife had died in the aftermath of childbirth. This brought his life to a state of personal and professional crisis. The work he was pursuing was incompatible with his obligation to his motherless children, and left him little time to pursue his true engineering enthusiasms. To cap it all, his partner in the steam-engine venture was in financial difficulties and it was increasingly obvious that the project was not going to prosper under his auspices.

In these circumstances, in 1774, Watt left Scotland to settle in Birmingham in the English Midlands and to pursue a new partnership with Matthew Boulton, the entrepreneurial 'hardware man' of that town.[7] Watt had met Boulton and members of the circle known as the 'Lunar Society'[8] some years before. In fact, one of that circle, Dr William Small, had given Watt important advice on drawing up the specification for his 1769 patent. Boulton, who had his own reasons for being interested in steam power, negotiated an agreement with Roebuck for the 'transfer' of Watt and his scheme to Soho, Birmingham. The incipient engine was removed from Kinneil and installed at the Soho manufactory. Watt's 1769 patent had already run a significant part of its course and the new partners' capacity to exploit it was only just reaching fruition, and so, as part of the new partnership arrangements, Boulton and Watt agreed to seek an extension of the patent by Act of Parliament. Here Boulton's well-known and well-honed lobbying skills came into their own and the partnership was granted an extension of the 1769 patent (redrawn in 1775 so as to constitute a new one) to the year 1800.[9]

In Birmingham, Watt certainly profited from and enjoyed the natural philosophical intercourse with his fellow 'Lunatics', as the members of the Lunar Society designated themselves. His natural philosophical knowledge and acumen increased considerably during this period. But the late 1770s and early 1780s were also a difficult time for Watt. The primary market for the first Boulton & Watt engines at this period lay in the mining districts of Cornwall where Newcomen engines had been used for many years in pumping the mines. Although in retrospect we can say that the Boulton & Watt engines offered decisive advantages in efficiency (notably in the saving of fuel, which was expensive in Cornwall) that ought to have made penetration of the market easy, at the time much resistance was encountered. The early engines were not problem-free, and canny Cornish mine venturers sought to drive hard bargains. The business of erecting engines and adapting them to local circumstances was vexed.[10] Watt had to spend much time 'on the road' in Cornwall attending to these matters, a task to which he was temperamentally unsuited. Contentious, face-to-face (or even toe-to-toe) negotiation, bargaining and industrial 'arm wrestling' of the sort that Boulton thrived upon, distressed Watt. Headaches, hypochondria and fits of depression plagued him – as he had previously put it, he would 'rather face a loaded cannon than settle an account or make a bargain'. Boulton's sometimes reckless business practices also worried Watt, who as a much more cautious businessman, was on occasion a valuable counterpoise to his partner's excesses of fiscal irresponsibility and blind enthusiasm.[11]

Gradually, a degree of business stability, together with staffing arrangements that removed Watt from the 'front line' of the Boulton & Watt venture, gave him more time and opportunity to pursue further steam experiments, other philosophical interests and inventive activities. The arrival of the Reverend Joseph Priestley in Birmingham in 1780 gave renewed impetus to Watt's chemical investigations and interests. Priestley was already renowned for his discoveries of new 'airs'.[12] He and Watt shared many chemical interests, but it was Priestley's experiments on the explosion of inflammable air with common air and with dephlogisticated air that led Watt into one of his few, high profile public appearances in the annals of late eighteenth-century natural philosophy. In 1784 Watt published papers in the *Philosophical Transactions* of the Royal Society of London, the most famous of them forming the basis of his claim to have been the discoverer of the compound nature of water.[13] This claim was far from uncontested either at the time or subsequently. In the *Philosophical Transactions* of the same year the eccentric, aristocratic natural philosopher Henry Cavendish published the paper that would be the basis for his own claim, and Lavoisier's later, but consummately theorized and prosecuted, experiments added him to the list of claimants.[14] What became known as the 'water controversy' did not create an enormous stir in the 1780s. Watt and his associates, for a time, vented their anger and frustration at what they saw as at best a careless appropriation of his ideas and at worst their outright theft. But there were good business reasons, if no other, for not creating too sustained a ruckus over the affair. It was not until the mid-nineteenth century that the water controversy was fought out in earnest. The protagonists then had their own agendas.[15] A crucial outcome of their struggle was to be a devaluation and mischaracterization of Watt's chemical work.

Even as he pursued these questions of high chemistry and philosophy Watt was also engaged in steam experiments that led him to the idea of 'expansive working' of the steam engine. This idea, and further key mechanical improvements, including the adaptation of the engine to rotatory motion, were the subject of Watt's second major patent of 1782. With rotatory motion now readily achievable through Watt's sun and planet gear system, the market for driving machinery was opened up to Boulton & Watt engines. The Albion Mill on the banks of the Thames housed one of the early rotative engines for grinding corn. Though not financially successful, and destroyed by fire in 1791, the Albion Mill was a valuable demonstration project, which helped the partnership to move into other rotative applications, especially in textiles, where the bulk of their engine business was to be.[16]

Another interest that Watt began to pursue at this time was in the bleaching of textiles. As part of a growing internationalization of their ventures Boulton and Watt visited Paris in 1786 in response to a government invitation to advise on the pumping machinery supplying water to the Palace of Versailles. They were also hoping, vainly as it turned out, to secure a privilege (the French equivalent of a patent) for their engines in France. Whilst in Paris, Watt, who had been elected a Fellow of the Royal Society of London in 1785, met many of France's leading natural philosophers. They were curious to meet a man whose steam engine improvements were already widely praised and whose papers in the *Philosophical Transactions* had created such a stir. These meetings were to greatly expand Watt's subsequent participation in the republic of letters and, specifically, in the chemical debates of the late eighteenth century. One of these meetings was with Claude-Louis Berthollet who revealed to Watt his work on the bleaching powers of dephlogisticated muriatic acid. On his return Watt conducted his own experiments and a long correspondence with Berthollet. He advised his father-in-law, who was in the bleaching business, about the new processes and, more generally, played a significant role in the introduction of those techniques into Britain.[17]

Mention of Watt's father-in-law reminds us that, after the loss of his first wife, Watt had married again, to Anne McGrigor. By all accounts a strong personality and a good organizer of Watt's household, Anne bore Watt two more children, Gregory (b. 1777) and Janet (known as Jessy, b. 1779). Meanwhile Watt's son from his first marriage, James Jr, underwent a modern education that included a period of study and travel on the Continent, partially supervised and guided by Watt's friend and chemical collaborator Jean André De Luc. Watt Jr was an independent-minded young man whose relations with his stepmother and his father were often stormy. He was apprenticed to the firm of Taylor and Maxwell in Manchester in 1788, where he participated both in the burgeoning scientific life of the town through the Literary and Philosophical Society (f. 1781) and in the radical politics of some of his associates. During a visit to Paris with Thomas Cooper in late 1791 and 1792 Watt Jr expressed solidarity with the early revolutionaries, a move which, when reported in the British press as the action of 'James Watt', caused his father much embarrassment and distress. For a while Watt Jr was in great danger in the increasingly violent factional maelstrom of the mature revolution. Eventually, in 1794, he managed to return to England and, after a period lying low to ensure that he would not be arrested by British authorities, Watt Jr returned to his family and to the business. From that point onwards Watt Jr was to assume an increasingly prominent role in the business affairs of the newly constituted Boulton, Watt & Co. as the business was gradually passed on to the next generation. Watt Jr, perhaps in an effort to assuage his guilt about early waywardness, became an assiduous, almost maniacal promoter and defender of his father's reputation, especially his claims as an inventor and discoverer.[18]

An early forum for Watt Jr's efforts in that direction was the patent trials of the 1790s. In the face of what they regarded as repeated infringements of their patents by engine 'pirates', Boulton & Watt had considered legal action from the early 1780s. But it was not until the early 1790s, when defaults on financial agreements by Cornish engine users became increasingly widespread, that they began to prosecute. A long, complex, expensive and vexed series of legal cases challenged Watt's ingenuity and stoked the fires of his hypochondria for most of that decade. The eventually successful outcome enabled Boulton & Watt to recoup a very large sum of money in arrears.[19] By this time, in 1800, the original patents had run their course and Watt retired from business.

Other domestic troubles stalked Watt in the early 1790s in the shape of the poor health of his younger children. Jessy succumbed to tuberculosis at the age of fifteen in 1794, after a harrowing illness. His parents invested great hopes in young Gregory. Bright and personable, he showed considerable promise as a philosopher, especially as a geologist, and studied at Glasgow College. But he too battled the dread disease and eventually died from it in 1804. The other child of Watt's first marriage to reach maturity, besides Watt Jr, his eldest child Margaret, had married James Miller, but she died in 1799, leaving four children. The Watts were no strangers to family tragedy.

The prospect of finding an effective treatment for Jessy was one of the major reasons why Watt entered into another of his key collaborations, this time with Dr Thomas Beddoes, on the production and use of 'airs' for medicinal purposes. Watt designed, and Boulton & Watt manufactured and sold, an apparatus for producing therapeutic airs and for delivering them effectively to patients.[20] But Watt's involvement went beyond this into both the chemistry of air production and of therapeutic effects. The steps leading to the establishment of Beddoes's short-lived Pneumatic Institution in Bristol are very revealing of Watt's ongoing defence of the chemistry of phlogiston. Along with his old friend Priestley, by now driven into exile in Pennsylvania by the Church and King riots in Birmingham in the early 1790s, Watt remained one of the last defenders of a phlogistic chemistry against the new chemistry most associated with the work of Lavoisier.

In his later years and in retirement Watt remained actively inventive, though now much of the hands-on work was done by his son or by assistants. With John Southern, Watt further developed the steam indicator, a device whose early forms he had devised in the late 1780s and early 1790s, and which became, in the nineteenth century, perhaps the most important of his lesser-known inventions because of its great utility in measuring the performance of steam engines.[21] Watt also enjoyed spending time in the attic workshop of his home, Heathfield Hall, where he developed machinery for copying statuary. He and his wife travelled, and also purchased land in Wales, establishing the estate that became known as Doldowlod where Watt and his surviving son pursued the habits of country gentlemen.[22]

Released from the cares of business, Watt enjoyed better health than he ever had during his most active years. He gained a reputation as a rather genial, avuncular sage, an ardent and entertaining conversationalist with a, perhaps surprising, taste for light fiction.[23] However, Watt, assisted and encouraged by Watt Jr and others of his circle, also assiduously attended to defending and cultivating his own public image and reputation. He monitored public accounts of his life and works and, where necessary, sought to correct or counter them. For example, when Olinthus Gregory, a professor of mathematics at the Royal Military Academy, Woolwich, with an eye for support of the underdog versus the 'establishment', included material favourable to the Watt steam engine rivals the Hornblowers in his treatise on mechanics, the Watts arranged for a damning review by John Playfair in the *Edinburgh Review* of just that part of the book.[24] Watt was under pressure from various quarters to give an account of the history of his inventions, and he agreed to do this in a roundabout way by correcting and annotating the articles 'Steam' and Steam-engine' that his friend John Robison had contributed to the third edition of the *Encyclopaedia Britannica* in 1797. Watt reviewed some of his earlier experiments on steam with the assistance of John Southern, and incorporated the results into his annotations. There too Watt published the last of a long line of accounts of his invention of the separate condenser.[25]

Watt received numerous honours, including Foreign membership of the French Institute in 1816 (one of only eight positions), and an honorary Doctorate of Laws from the University of Glasgow in 1806. He reportedly turned down a baronetcy when sounded out on the question by the government through Sir Joseph Banks. Had this been accepted, or publicly known, it would have threatened the 'outsider' status that the Watt camp often sought to cultivate, partly through complaints about the *failure* of government to honour the great man with an appropriate title! Watt also made a few benefactions, including a gift of books in 1816 to his native town of Greenock. Eventually his son was to fund a library building in which to house those books, which became the Watt Monument Library, and a focus of local commemorations of the town's most famous son.[26]

Watt died, after a short illness, aged 83, on 25 August 1819 at Heathfield Hall. He was buried in Handsworth Parish Church in Birmingham where his business partner Matthew Boulton had preceded him. A statue in marble by Francis Chantrey soon dominated a side chapel in which Watt was buried. The obituaries flowed, the most famous and oft-quoted being that in *The Scotsman* newspaper by Francis Jeffrey, the founder of the *Edinburgh Review*. Though Watt was technically the improver of the steam engine, Jeffrey said, 'he should rather be described as its *inventor*'. As such he had conferred inestimable benefit upon his country and upon humanity. He had 'armed the feeble hand of man ... with a power to which no limits can be assigned ... and laid a sure foundation for all those future miracles of mechanical power which are to aid and reward the labours of after generations'.[27]

The making of the legend was already well under way when the 'Great Steamer' departed, but there was a lot more to come.

The Structure and Strategy of this Study

This book has both a general and a specific intent. The specific purpose is to partially recast our understanding of James Watt's improvements of the steam engine and the natural philosophical basis for them. The origins of the steam age, I will argue, lay in an intellectual universe that is profoundly alien to our own. I do not claim to have fully recovered that intellectual universe, but will be happy if I have pointed in the right general direction. So the comprehensive history of Watt's chemical connections and activities through what we can loosely call the 'Chemical Revolution' remains to be written. This is not it. Indeed part of my specific purpose is to execute a necessary move preliminary to such a comprehensive chemical reconstruction – to suggest that a history of Watt's chemistry, if it were to be written apart from what are conventionally seen as his 'steam engineering' activities, would reinforce an artificial divide between the two which hinders the understanding of both. The broader purpose of the book is to exemplify the ways in which historical sources can be profoundly misleading for such attempts at reconstruction. In doing this I am pursuing old cautions about the perils of Whig history, and clearing away historiographical undergrowth by showing the contingent purposes of documents that have often been taken as transparent historical sources.

The book is divided into two main sections – 'Representations' and 'Realities'. Of course, I am aware that the realities that I offer in the second part of the book are simply my representations of the realities as I see and reconstruct them. My hope is that others will find them a persuasive, or at least a suggestive, account, not that they will be treated as the end of history so far as my subject is concerned.

In 'Representations' I pursue some of the sculptural, pictorial and textual representations of James Watt between his own time and the present, with emphasis on the nineteenth century. It was during the nineteenth century that Watt became a powerful and protean symbol. I am conscious, in presenting these representations, of the splendid work done recently on representations of Watt by Christine MacLeod. Professor MacLeod's work has been an important inspiration and substantive source for what follows.[28] However, we intersect rather than coincide. MacLeod's focus is that of an economic and cultural historian. She looks from representations of Watt outwards into the British economy and society in the nineteenth century and is concerned primarily with how images of Watt are constructed in the service of broader social and economic purposes. I look, primarily, from images of Watt 'inwards' into the knowledge economy, or, perhaps better, knowledge 'ecology'. As an important contributor to the development of arguably the most important technology of industrialization Watt became a symbol that many groups in the knowledge business were anxious to claim. Whether he could, or should, be characterized as a 'natural philosopher', an 'engineer', a 'philosopher engineer', a 'craftsman', or a chemist, as a discoverer or as an inventor, was an issue much contested. A particular strength of Professor MacLeod's research on Watt's reputation is the picture she draws of its construction by the ranks of practical mechanics and other workers. This is an aspect that I am conscious of neglecting in the following pages, partly because my focus is upon those learned and professional classes in whose hands the construction of Watt's learned character primarily lay.

In analyzing these varied representations I draw a distinction between the 'mechanical' and the 'chemical' Watt. My reason for doing so is that my claim in 'Realities' is that Watt was in a crucially important sense a chemist. This is an unusual, and far from obviously justified, claim – at least in the form that I advance it. Thus, in dealing with representations, I need to examine how and why the mechanical Watt obscured the chemical. Accordingly, Chapter 1 explores representations of Watt in marble, stone and brass, the famous story of Watt and the kettle, and the much less famous story of Watt and the steam indicator, by way of illustrating the dominance of the mechanical Watt in public representations of him. In Chapter 2 I offer some suggestions as to why the chemical Watt disappeared in the course of the nineteenth century. One reason was that the man himself, and his close associates, chose to recast the nature of his achievement in a way that 'proofed' it against rapidly shifting understandings of the nature of heat. Another was that the chemical claim to fame that was allowed to survive this process of self-fashioning, namely Watt as candidate discoverer of the compound nature of water, became the subject of a substantial controversy in which, for different reasons, both the supporters and opponents of Watt's chemical pretensions helped to obscure his chemical understandings. In Chapter 3 I explore the creation of Watt the engineer during the same period. Some advocates of the development of 'engineering science', in particular, adopted Watt as a symbol of what they held dear. In so doing there was an understandable tendency to modernize the man and his ideas. Overall then, characterizations of the great man are related to his self-fashioning, to the commemorations pushed by family and friends, and to the concerns and ambitions of groups seeking, or indeed declining, to adopt Watt as a symbol. The outcome of this 'orgy' of contending representations was a very useful, but profoundly misleading, 'mechanical' Watt.

In Part 2, 'Realities', I turn to the reconstruction of the chemical Watt. Chapter 4 seeks to portray his chemical world and how it came to be that way. Watt was not 'trained' as a chemist in any meaningful sense. But through a series of informal associations, with Joseph Black, J. A. De Luc and Joseph Priestley in particular, and by a great deal of self-education, Watt did become a figure to be reckoned with in the chemical world of the late eighteenth century. I argue that his chemical ideas were the product of the interaction of theory and practice and in so doing bring together episodes in his life and work that are usually considered to be unrelated to each other. The fact that those episodes can be brought together in this way is further evidence, I contend, for the importance of the chemical thread in his work. Having depicted Watt's development as a chemist, I then turn in Chapter 5 to advance and argue the claim that the steam-engine improvements were a product of Watt's chemical philosophy and that in that, and other, senses the steam engine of the late eighteenth century ought to be seen as a chemical device. The final chapter offers a concluding case study of Watt and the steam indicator. Watt's development of the early versions of this device has encouraged some to regard him as prescient about the science of energy, even as what we might call a 'proto-thermodynamicist', who anticipated some of the most central and important physical insights of the nineteenth century. In showing how elisions have been made between Watt's use of this device, and the use made of it in the theory of thermodynamics, I suggest that his understanding of the indicator was essentially chemical in nature and fundamentally at odds with, rather than being an anticipation of, nineteenth-century thermodynamics.

In ten years time we will mark the 200th anniversary of Watt's death. The 100th Anniversary in 1919 recognized only the mechanical Watt. Perhaps the next commemoration will do more justice to the complexity of the remarkable achievements of a man who was, I suggest, a stranger to our usual ways of understanding him.

1 OF STATUES, KETTLES AND INDICATORS: · THE 'MECHANICAL WATT'

I begin with a tale of three representations of James Watt. The first is a famous and much-travelled marble statue, sculpted by Francis Chantrey. As I write it looms in the entrance hall of the National Portrait Gallery in Edinburgh. Soon it may move to its intended permanent location at Heriot-Watt University. It began life in Westminster Abbey as an unofficial national commemoration of a great national hero and benefactor. Though some considered its scale inappropriate, others thought its monumentalism in proportion to the significance of the man and his deeds. In the twentieth century its detractors eventually won and in 1960 it was replaced in the Abbey by a small bust of Watt and moved to the Transport Museum at Clapham. Its passage was surely barely eased by strained arguments that Watt should take a place of honour among the paraphernalia of steam locomotion. When the Transport Museum was relocated to

Figure 1.1. Watt by Chantrey in Suspension, c. 1974. Originally in Westminster Abbey, the statue was left behind when the Transport Museum, where it had been housed from 1960, moved from Clapham to York. (U.K National Archives, Author's Photo)

York, however, the attractions of the giant statue had faded and for some time it stood alone in the empty sheds at Clapham while its carers sought it a home (See Figure 1.1). The Science Museum did not want it; the stationmaster of Glasgow Railway Station was unimpressed by the argument that it would grace the concourse there. The monumentalism of the nineteenth century was the source of mounting embarrassment to the twentieth until St. Paul's Cathedral agreed to take it, providing someone else would fund the reinforcement of the floor where the marble edifice was to stand! From St. Paul's the statue made its way to Edinburgh and there it stands in the entrance hall of the National Portrait Gallery, barely accommodated even without its original pedestal. The absence of the pedestal means that visitors have to make do with a photograph of the lapidary inscription by Lord Brougham that adorned it. Little noticed by passing patrons of the Gallery is the geometrical pattern that Chantrey had carefully engraved into the scroll on Watt's lap. It symbolized the work that Chantrey believed Watt should be remembered for most.[1]

For a while within recent memory a large yellow kettle stood on Gourock Pier not far from the site of Watt's birth in Greenock. It functioned then as a tourist information centre but had been constructed originally for a Glasgow garden festival. If Glasgow ever hosts an Olympic Games perhaps the kettle will be the equivalent of the Victa lawn mower and the Hills Rotary Hoist that were represented at the opening ceremony of the 2000 Sydney Games. Glasgow's, or at least Greenock's, most famous son is forever associated with the kettle that supposedly gave the young Watt the inspiration for his improvements of the steam engine. Innumerable juxtapositions of Watt and the kettle litter the cultural landscape. Most derive from two nineteenth-century paintings, one by Robert Buss, the other by Marcus Stone, which picture an anecdote concerning Watt's youth that had been unearthed and propagated by his son James Watt Jr. Like Chantrey's statue the kettle story has hidden aspects. It is a 'true myth'. When properly understood the association of Watt and the kettle teaches us much about the origins of the steam age.

The third representation of Watt with which I begin my tale is also in Greenock, on or near the site of Watt's birth. This is a bronze, full-length statue of the great man, which looks out over the Clyde. It was unveiled on 1 June 1908 by the American-Scottish steel magnate Andrew Carnegie when he opened the Watt Memorial Engineering and Navigation School on the site. Modern sightseers must often puzzle about the strange object that Watt is thrusting forward in his right hand. But among those applauding Carnegie on that spring day in 1908 many would have understood that object very well from their everyday experience of operating and tending the marine steam engines which powered many of the ships on the river behind them. It was an 'indicator', a device used to measure the performance of steam engines, to diagnose problems with their operation and so enable adjustments to be made and their efficiency to be improved. So Watt, on the Clyde, wielded a device supposedly of his invention that had particular meaning for his local audience and for the engineering students who were to pass through the School founded in his name – a fitting commemoration. However, we will see that the indicator, too, has other hidden aspects and meanings.

The statue, the kettle and the indicator are nineteenth-century constructions of Watt, which embody a mythology about the great man. This book is about the construction of that mythology, the purposes that it served, how it obstructs our understanding of the intellectual origins of the steam age, and how we can see 'behind' it. The image of Watt to which the statue, the kettle story and at least one version of the indicator story all contributed was of a scientifically informed, or 'philosophical' engineer. Watt had investigated the physics of heat, and had anticipated some of the crucial features of thermodynamics as developed in the nineteenth century. On the basis of this understanding he had worked out how to improve the performance of the steam engine and, through the indicator, how to monitor that performance. His genius for contrivance had produced many other inventions and improvements such as the sun and planet gear system and the 'governor', but understanding of the physical properties of steam was seen to be at the centre of his achievement.

While this interpretation of Watt and his work is coherent and its origins understandable, I contend that it fails to capture the eighteenth-century realities of Watt's work and how he understood steam and the engine improvements that eventually made his fortune and his name. The science of heat was certainly central to all that Watt did, but that science was chemical in nature. Strange as it may seem to us, for him the steam engine was at its heart a chemical device. Let's discuss in turn, and in more detail, the statue, the kettle and the indicator to see how they reflect the constructions of Watt in the nineteenth century.

Where the Statue Stood

The statue by Chantrey that began its wanderings in Westminster Abbey was, in fact, the third of five versions of the same basic, seated model that were distributed around Britain. The others were in Handsworth Parish Church (where Watt is buried); in the Watt Library, Greenock; in George Square, Glasgow; and in the Hunterian Museum in Glasgow. This sculpture was widely regarded as one of the finest examples of the work of Francis Leggatt Chantrey. The image shared many features of early nineteenth-century neoclassicism, but in other respects was vaunted – not least by Chantrey himself and his publicists – as a new, classic, national style of sculptural portraiture that avoided 'allegorical aids and classicising fictions such as partial nudity and antique drapery' yet effectively conveyed by more direct means the character and achievements of the subject.[2]

Chantrey's early sculptures concentrated on busts of scientists, engineers and doctors. Certainly in the case of Watt his work became the focus of debate about which sorts of national heroes should be commemorated in public statuary. Many, including the promoters of the Watt subscription, felt that it was vital to celebrate the progenitors of the productive forces transforming the world of enterprise and industry as much as the statesmen and military figures who were more traditionally accorded such treatment.[3] In this regard the installation of Watt in Westminster Abbey had particular significance. Some observers may even have seen the damage caused by the very process of installing the statue in St Paul's Chapel in the Abbey as symbolic of a changing of the guard.[4] In the mid-nineteenth century, Hugh Miller recorded an encounter with the statue:

> I was a good deal struck, in one of the old chapels, – a little gloomy place, filled with antique regalities sorely faded, and middle-age glories waxed dim, – by stumbling, very unexpectedly, on a noble statue of James Watt. The profoundly contemplative countenance – so happily described by Arago as a very personification of abstract thought – contrasted strongly with the chivalric baubles and meaningless countenances on the surroundings tombs. The new and the old governing forces – the waxing and the waning powers – seemed appropriately typified in that little twilight chapel.[5]

Here Miller puts Watt (as represented by Chantrey) forward as a symbol of all that is new and rational in the modern world, and contrasts his monument with the meaningless regalia of the 'old governing forces'. The reference to Arago's description of the statue as representing abstract thought so well is significant.[6] Chantrey was widely noted for his powers of representing mind, an aspect of his naturalism. The idea that he had somehow captured Watt's mind became part of the standard repertoire of commentary on the Watt statue.[7]

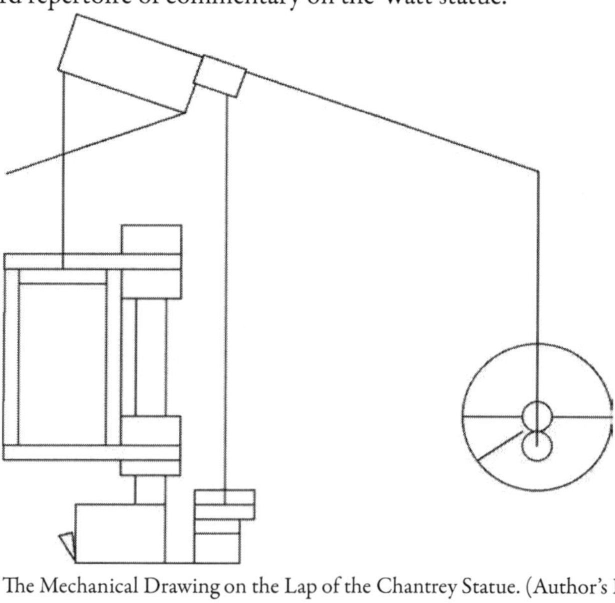

Figure 1.2. The Mechanical Drawing on the Lap of the Chantrey Statue. (Author's Drawing)

Perhaps less noticed or discussed was the imagery on the scroll on Watt's lap, over which his hand, with its dividers, and his noble head, were poised. There Chantrey carefully depicted, in another gesture towards naturalism, the mechanical contrivances for which Watt was to be remembered (See Figure 1.2). The accurate representation of Watt's mechanical ingenuity was part of the naturalism of the sculpture. That ingenuity was an emanation of the mind captured in the countenance. The 'mechanical' is represented not by a model or a physical device but by a geometrical drawing of it. Watt, like Newton, is depicted as a geometer and therefore a philosopher. His geometry is that of the spatial arrangement of components of the steam engine with the condenser separate from the cylinder, and the sun and planet gear arrangement solving the problem of producing rotative from reciprocating motion.[8]

The Chantrey statue of Watt in Westminster Abbey was the result of a campaign run by Watt Jr and a number of his friends, including Charles Hampden Turner, which culminated at a subscription meeting in the Freemason's Hall on 18 June 1824. As MacLeod explains, this was a highly symbolic occasion so far as the contest over selection of national heroes was concerned. Many worthies spoke on that occasion and the words have come down to us. One, Sir Humphry Davy, President of the Royal Society, spoke of Watt's chemical credentials but most were anxious merely to emphasize that Watt was no ordinary engineer but also a philosopher.[9] The most famous, and enduring, words associated with the statue, however, were those penned by Watt's friend and supporter, Henry, Lord Brougham for the inscribed base of the edifice:

> Not to perpetuate a name
> Which must endure while the peaceful arts flourish
> But to shew
> That Mankind have learnt to honour those
> Who best deserve their gratitude
> The King
> His Ministers and many of the Nobles
> And Commoners of the Realm
> Raised this monument to
> JAMES WATT
> Who directing the force of an original genius
> Early exercised in philosophic research
> To the improvement of
> The Steam Engine
> Enlarged the resources of his country
> Increased the power of man
> And rose to an eminent place
> Among the most illustrious followers of science
> And the real benefactors of the World
> Born at Greenock MDCCXXXVI
> Died at Heathfield in Staffordshire, MDCCCXIX.[10]

These words were most carefully crafted and negotiated between Brougham and Watt Jr. Brougham had originally used the phrase 'early trained in philosophic research' but Watt Jr had objected strenuously. The not-so-deep subtext in this exchange, was the nature of the relationship between Watt and Joseph Black and the extent to which Watt had learned from the professor. Watt Jr was anxious to ensure that his father's autonomous philosophical character was clearly acknowledged.[11] At one point, Watt Jr asked Brougham 'whether it would be proper in the Inscription to allude to' his father's discovery of the composition of water. Clearly, Brougham thought it a bad idea, and Watt Jr did not insist. Had such a reference been included then the chemical Watt might have gained a little more iconographic prominence.[12] Unfortunately we do not have Brougham's reasoned response to the suggestion. Perhaps he thought it best to remain on secure ground, or perhaps his rejection of the idea rested on the maintenance of harmony between inscription and sculpture. The latter so clearly depicted Watt as geometer and mechanician, that to introduce chemical discovery into the inscription would be confusing and detract from its coherence. It is worth making the obvious point here that the association of Watt with *mechanical* contrivance, which is so common in imagery of him, reinforces the mechanical rather than the chemical Watt. Not the least of the advantages of the mechanical Watt was the ease with which he could be depicted. The chemistry of steam represented more of a challenge in that regard. There the imagery of the kettle was, as we shall see, a possible, but highly problematic, solution.[13] For the moment we can say that Brougham's inscription, the dividers in Watt's hand and the mechanism depicted on the paper on the statue's lap all contributed to Watt's status as philosopher, geometer and mechanician being set in marble.

In the later nineteenth century a number of other sculptors were commissioned to produce statues of Watt. Some of these were effectively imitations of Chantrey's statue. Even those not directly imitative often depicted the dividers and scroll, thus perpetuating the head and hand image of the philosopher-engineer. This is the form, for example, of the statue by William Theed erected in 1857 in Manchester, and also of that by Henry Charles Fehr in City Square, Leeds.[14]

There does appear to be a tendency for later nineteenth-century statues to depict Watt with specific devices of his invention. The statue by Charles Benham Grassby, now outside the 'People's Palace' on Glasgow Green, shows him leaning on part of an engine, though still holding a pair of dividers (See Figure 1.3).[15] A similar statue, in that it features Watt standing beside a cylinder or condenser, was commissioned from the sculptor Albert Hemstock Lodge as part of the 1905–8 extension scheme of the Port of Glasgow. It still stands today at 16 Robertson Street, Glasgow at the right of the pediment of the Clydeport building.

The less papyrophilic, more device-centred, depiction of Watt was in tune with the growing tendency to describe him as a craftsman or craftsman-engineer in the

Figure 1.3. Watt by C.B. Grassby outside the 'People's Palace', Glasgow Green. (Author's Photo)

later nineteenth and early twentieth century. It had appeared earlier in statues supported or erected by artisans, including in the very early statue commissioned by the Glasgow Mechanics' Institute in 1824. This work, by John Greenshields, has a life-size Watt leaning on a steam cylinder, and was imitated by Grassby. Ray McKenzie has remarked on the 'populist egalitarianism' that engendered the mechanics' statue of the 'Great Steamer'. That populism was reflected in its form. Later in the century, though, even the statue of Watt in the courtyard of the Oxford Museum of Natural History, dating from 1860, though it features more classical drapery and the dividers, has a bet each way when it also depicts Watt holding a device.[16]

An interesting variant, which requires a whole story to itself, and which I will return to below, is the statue by H. C. Fehr in Greenock which was unveiled by Carnegie in 1908. As already discussed, this too involves a device. The statue is in fact a copy of the one that Fehr executed for Leeds City Council in 1898. However, instead of the dividers in the right hand of the Leeds statue, in Greenock Watt holds up an indicator. The move to standing (active) figures in modern dress, holding or leaning upon actual devices – rather than contemplating geometrical diagrams with large mind – is, I think, a significant one. Watt the craftsman-engineer was being celebrated in these later statues by contrast with the philosopher-engineer of the earlier ones. What is certain is that all these figures, whatever their differences and whether by design or by default, set the 'mechanical Watt' in stone or brass.

Kettle Mythology

Watt and his kettle are everywhere still, today. Figure 1.4 depicts the large yellow kettle already mentioned. This was built at the James Adam's ship repair yard in Gourock in 1988 before being transported by barge up the Clyde to be installed at the Glasgow Garden Festival. The kettle was sponsored by the Solid Fuel Advisory Service under the slogan 'We're Steaming Ahead'. Local worthies, speaking at its 'launch', expressed hope that the economic tide was turning for the Inverclyde district and saw the kettle as symbolizing the 'natural imagination and inventiveness for so long associated with the local people'. The *Greenock Telegraph* elaborated: 'Inspired by Greenock-born engineer and steam pioneer James Watt, the Inverclyde Kettle symbolizes the area's spirit of enterprise and challenge through the sugar trade, ship building and heavy engineering industries to modern day electronics'. At the Garden Festival the kettle sat on a bed of real coal adorned by flowers simulating flames, whilst inside exhibitions and videos promoting Inverclyde could be seen by visitors. The kettle was subsequently

Figure 1.4. The Big Yellow Kettle: Watt Mythology Lives. The Kettle is 'Launched' in 1988 for Transport down the Clyde. Built for the Solid Fuel Advisory Service and Carrying the Slogan 'We're Steaming Ahead.' (Photograph by Courtesy of *Greenock Telegraph*)

paraded through the streets of Greenock on its way back to Gourock Pier where it spent some time as a Tourist Information Bureau.[17]

The depth and range of the kettle mythology can be illustrated by the rather mature looking Watt and his kettle found in the late nineteenth-century Japanese wood block print reproduced in Figure 1.5. Watt, overseen by his aunt, is described as 'measuring water drop by drop as it condensed from steam coming out of the spout of a kettle in an attempt to produce steam-powered locomotion'. Although his aunt scolded him for wasting time, we are told that he nevertheless 'developed the steam engine and rendered meritorious services'.[18] This very attractive woodblock print was one of a series distributed by the Japanese Ministry of Education during the early Meiji period. It dates from about 1870. This was a time when Japan was opened up to Western science and technology, and evidently to its mythologies also. Keiko Suzuki explains that the message of such 'enlightenment pictures' (*Kaika-e*) was about persistence and of difficulties overcome. They presented symbols of Western culture. Together with writings of the time they blended Smilesian self-help and utility 'with traditional Confucian stoicism and Samurai self-reliance' as a role model for Japanese youth.[19] Whether in Greenock or in Tokyo, the urging of industrial development and tales of Watt's kettle are never far apart.

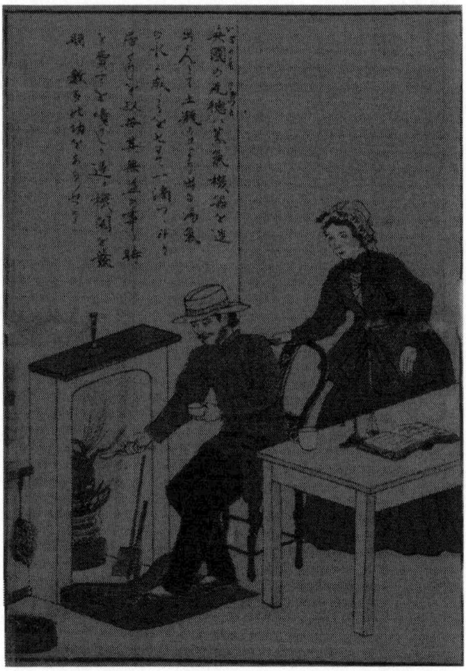

Figure. 1.5. Watt's Kettle in Meiji Japan (Reproduced, with permission, from a woodblock print in the collection of Yokogawa Archives, Yokogawa Electric Corporation).

Returning to recent times, I find that Watt and the kettle are a popular reference point in sermons, often coupled with Robert the Bruce and his spider or Newton and his apple. One such sermon, delivered in Arkansas in 2001, quoted from a book by William Barclay, *The Gospel of Luke*: '"Many people saw steam raise the lid of a kettle; only James Watt went on to think of a steam engine. Many people saw an apple fall; only Isaac Newton went on to think of the law of gravity"'. The preacher continued: 'We might add that many people saw a shepherd return home with a lost sheep draped over his shoulders, but only Jesus went on to think of God's loving providence.'[20] Not only God but also Mammon has been served by the story. It has been widely used in advertising, on one occasion in developing a campaign for the company LG. The advertisement is about seeing things differently on the 'Flatron' flat-screen television just as Watt had seen things differently when he looked at the steaming kettle. The words in the steam ask, 'Do you see a revolution here?' The main text of the advertisement responds: 'Revolution in a kettle? Yes. James Watt saw it. He saw the steam engine in it. And laid the foundation for the great Industrial Revolution. A little extra zeal to see things differently, and a revolution is on hand'. Finally we are advised: 'Expand your life'. Over the years Watt and the kettle have helped to sell, among many things, whisky, nuclear energy and, perhaps more predictably, steam boilers.[21]

Similar evangelical fervour characterizes management literature on innovation and its injunctions to 'think outside the square'. Richard Farson, in his aptly titled book *Management of the Absurd* (1996) describes how James Watt supposedly saw something that millions of other people had also seen but 'not seen'. It was, Farson claims, Watt's observation of how steam coming from his tea kettle could be used to power steam engines that sparked the Industrial Revolution. Watt also saw the 'invisible obvious' that so many others could not.[22] I am not, needless to say, endorsing the interpretations here, but simply illustrating the ubiquity of the myth and the fact that the kettle inspiration is so often depicted as somehow giving Watt the idea of steam power *in toto*.

My favourite piece of secular literature that invokes Watt and the kettle is the 'Penguins' sketch in Episode 38 of *Monty Python's Flying Circus*. It begins with the words of the presenter, played by John Cleese, whose manic hyperbole, progressively contradicting itself, captures beautifully our cultural confusion over the role of accident, genius and hard work in discovery and invention. However, when Cleese comes to the Watt anecdote we are told: 'James Watt watched an ordinary household kettle boiling and conceived the potentiality of steam power'.[23] Even the most literate and, in other respects, perceptive of modern popular invocations of the myth treat it as being about Watt's discovery of the *power* of steam.

These few examples of the deployment of the myth of Watt and the kettle show that the story is ubiquitous, that it has a number of variants and that, most commonly, it simply makes the connection between steam and power. Its telling often involves the old canards that Watt invented the steam engine or that he was the first to see the potential of steam power. The story is sometimes told with great earnestness. On other occasions it is recounted with a view to deflating or ridiculing it. The myth is modified creatively, it is perpetuated, and it is 'busted'. In fact, busting the myth is very popular. Only clerics, management gurus and comedians seem to take it seriously as a piece of history. Works for juveniles are sometimes very careful in using the story. Neil Champion's biography of Watt for young people, for example, tells the tale and reproduces a nineteenth-century painting by Marcus Stone depicting the young Watt contemplating steam (of which more in a moment). Champion warns, however, that the story may not be true![24] So where did the myth come from? Let us look at its history in both word and image.

On 14 May 1845, James Patrick Muirhead wrote to James Watt Jr from Hanover Square in London describing a painting that he had just seen in an exhibition at the Royal Academy of Arts. The painting, number 673 in the catalogue of the exhibition, was by Robert W. Buss and titled 'Watt's First Experiment on Steam'. Muirhead noted that the catalogue cited a passage from François Arago's *Eloge* of James Watt 'in which the anecdote is given of his watching, in the presence of his Aunt Miss Muirhead the escape and condensation of steam from a tea kettle'.[25]

Figure 1.6. R.W. Buss, 'Watt's First Experiment on Steam'. Engraved by James Scott. (Copyright ©. Reproduced with permission from the Thomas Ross Collection www.rosscollection.co.uk)

Muirhead was himself of the same family as the mentioned 'Aunt Miss Muirhead'. He was descended from her brother, whilst her sister was the mother of James Watt. Muirhead, who gave up a legal career and became in effect a literary *aide-de-camp* to Watt Jr, had translated Arago's *Eloge* of Watt into English in 1839. There he rendered Arago's telling of the anecdote of the kettle as follows:

> Sitting one evening with his aunt, Mrs. Muirhead, at the tea table, she said, 'James Watt, I never saw such an idle boy; take a book or employ yourself usefully; for the last hour you have not spoken one word, but taken off the lid of that kettle and put it on again, holding now a cup and now a silver spoon over the steam, watching how it rises from the spout, and catching and connecting the drops it falls into, are you not ashamed of spending your time this way?'... Mrs. Muirhead's reproof will present itself to our minds under a totally different aspect; the little James before the tea-kettle, becomes the mighty engineer preluding to the discoveries which were to immortalise him...[26]

An engraving of the picture that Buss exhibited that day is reproduced in Figure 1.6.[27] As Muirhead noted, the picture is clearly based on the passage from Arago's *Eloge* of Watt: the personnel involved are Watt and his aunt; the young Watt has his hand on the tea kettle as if performing some of the operations that Arago described, and his aunt is wielding a watch measuring the time that she thinks Watt is wasting. Note, however, that in itself the Buss painting does not reflect that part of the anecdote concerning the *condensation* of steam. The cup and the spoon are not being wielded, though they do lie there on the table. The painting is most likely to be read as showing the young Watt discerning the power, not the properties, of steam. Having seen the painting, Muirhead's immediate reaction was amusement, as he described it to Watt Jr:

> The picture is not so bad as we thought it likely to be, and I even heard it called 'very clever' by one critic. James Watt you are to know was then a good looking boy in a purple velvet coat (!) and Miss Muirhead a stiff and stately old dame with a very pretty tea:service of Japan china and an old fashioned watch on which she indignantly pointed out the time that her nephew had wasted.

Muirhead was evidently amused by the sartorial and domestic licence taken by the artist. Nevertheless he found a 'distant resemblance' between Miss Muirhead as depicted and the portrait of her sister (Watt Jr's grandmother) that hung in Watt Jr's dining room, 'although, as was natural to expect', Muirhead added rather ungallantly, 'the married sister greatly excels the unmarried one in attractions'.[28] Later, in his *Life of Watt*, published in 1858, Muirhead recounted the anecdote again, and there, for reasons we shall shortly examine, he vouched for its accuracy.[29] For him this anecdote was no myth but a reliable account of a real incident.

In 1863, some twenty years after the exhibition of Buss's painting, another artwork was executed, 'Watt Discovers the Condensation of Steam' by Marcus Stone.[30] This rather attractive genre painting has been widely reproduced (see Figure 1.7). It represents a very different scene from that painted by Buss. Young Watt is still playing with the kettle, but this time he is pictured holding a spoon in the steam rather than fiddling with the lid. The scene is not at his aunt's house but rather in his own home. The young Watt is seated at the table with his mother and father. The boy who has left the table and gone to serve a customer, is almost certainly James's brother John. Like his father, John inhabited the marine and mercantile world, as represented in the shopfront annexed to the domestic quarters. John died at sea at the age of twenty-three. In contrast with this active world, and with the drudgery of the maid who also has her back to him, Watt the child occupies a contemplative space in his reverie on steam. The parents' attention is on their strange son. They appear fascinated yet perplexed by their offspring, fixed in an attitude of uncomprehending devotion. Gone is the scolding aunt. The adults here, far from reproving the child for wasting time, appear to be admiring his genius. It is also notable that whilst the young Watt in Buss's painting is plausibly an errant teenager, the Watt in Stone's version is younger, almost cherubic. He is innocently open to the inspiration coming his way.[31] Many subsequent renderings of the myth of the kettle derive from Stone's painting insofar as they depict Watt at home with his mother. The painting also seems to have inspired a play for young people published in 1936, at the time of the Watt bicentenary, called 'Letting off Steam'.[32] This is a charming little drama, set around the Watt family hearth. It was undoubtedly performed in schools in the UK to mark the 200th anniversary of Watt's birth.

Figure 1.7. Marcus Stone, 'Watt Discovers the Condensation of Steam'. Engraving by James Scott. [Reproduced from A. L. Baldry, *Marcus Stone, R. A. His Life and Work* (London, 1896), p. 13.]

The images are seductive, but let us return to the words. So far we have traced the myth back to François Arago's *Eloge of James Watt* published in 1839. Where did Arago get the story? In fact he obtained it from James Watt Jr. In 1834, Arago visited Britain to attend the meeting of the British Association for the Advancement of Science in Edinburgh. His other prime purpose during the trip was to collect material for the *Eloge* of Watt that he was due to deliver to the Académie des Sciences at the end of that year. Thus began a long correspondence and a series of meetings between Arago and James Watt Jr about the *Eloge* that lasted until its long-delayed publication in 1839 and beyond.[33] The story of Watt and the kettle was one among many pieces of information that Watt Jr communicated to Arago in the wake of his visit. A version of the kettle story was sent to Arago first in late October 1834 in a copy of a letter received by Watt Jr from Dr James Gibson. Gibson, who was married to Agnes, a daughter of Watt Jr's sister Margaret, had received anecdotes of the young Watt from a Miss Jane Campbell, of whom more in a moment. He knew that Watt Jr was seeking material on his father's early life to send to Arago. Gibson's letter told the story thus:

> When a boy at school he [Watt] had not robust health whereupon the School Master recommended him to be withdrawn for a time & sent to the country to recover his health – while the elders were at table the subject of this *éloge* was found sitting by the parlour fire holding a plate (which he had filched off the table) to the spout of the boiling tea kettle noticing carefully the distances at which the steam condensed when it came in contact with the cold plate – who can say but this gave him first the idea of the power & elasticity of steam? – This is not exactly the way Miss Jane Campbell told the story but the facts are the same.[34]

This was probably the version of the story that Arago used in his spoken *éloge*, delivered to the Académie in December 1834.[35]

Watt Jr wrote directly to the source of these anecdotes, Miss Jane Campbell, and learned that she had received them from her mother, Mrs Marion Campbell, a cousin of James Watt who had spent much time with him as a child. In fact the anecdotes had been dictated to Jane Campbell by her mother in 1798. Miss Campbell now sent a copy to James Watt Jr for Arago's information. So it was that in January 1835 Watt Jr wrote first to Arago's 'minder' J. B. Pentland and then to Arago himself with news of this material. Watt Jr explained to Pentland that the anecdotes had been written down some time ago. Although Miss Campbell had often told them to him, he was unaware until recently that they had been written down so early: 'from what I know from other sources, I can vouch for the authenticity of every word of it. It shows my father to have been ab incepto the same in talent & disposition, he continued to the end of his life. Mr Arago cannot do better in revising that part of his memoir, than to adopt every word of it in a literal translation'.[36] A few days later Watt Jr wrote to Arago sending a copy of Miss Campbell's Memoranda. Watt Jr was again anxious to emphasize the authenticity of the material:

> Mrs Campbell was first cousin to my father, being a daughter of his Mother's brother, Mr Muirhead, who was a builder and timber merchant in Glasgow. She was nearly my father's age, and they were brought up a good deal together at the houses of their respective parents. I knew her in after life, as a person of strong sense and clear judgement and possessing an affectionate esteem for my father. She was fully aware of the High estimation he had then attained, and it now appears she committed to her daughter's care this record of her recollections of his early years, which probably even at that time (in 1798), she was the only person capable of giving.[37]

Once again Watt Jr urged Arago to use the entire 'authentic' narrative in place of the earlier version.

Let me focus now on precisely what was received from Miss Campbell and what was sent to Arago and used in other printed recountings of the story. There is some difficulty is consulting the communications between Watt Jr and his cousin Miss Jane Campbell for the simple reason that in March 2003 the manuscripts were sold at Sotheby's in London and their current whereabouts are unknown. Fortunately the catalogue of the sale gives us considerable detail about the 'kettle documents' including a facsimile of part of the original Memoranda from Miss Campbell to Watt Jr.[38] Here, unsurprisingly, we find the central element of the story of the kettle and the scolding aunt and also a brief reflection upon the story. On examining this document closely, however, it became apparent that the original words, in Miss Campbell's hand, had been edited, if ever so slightly (see Figure 1.8). The word 'power' had been struck through and the word 'properties' written above it. In transcribing the manuscript, the Sotheby's catalogue quite happily uses 'properties', as if the correction had simply set right a slip of a single pen. I believe, however, that we are dealing with more than a

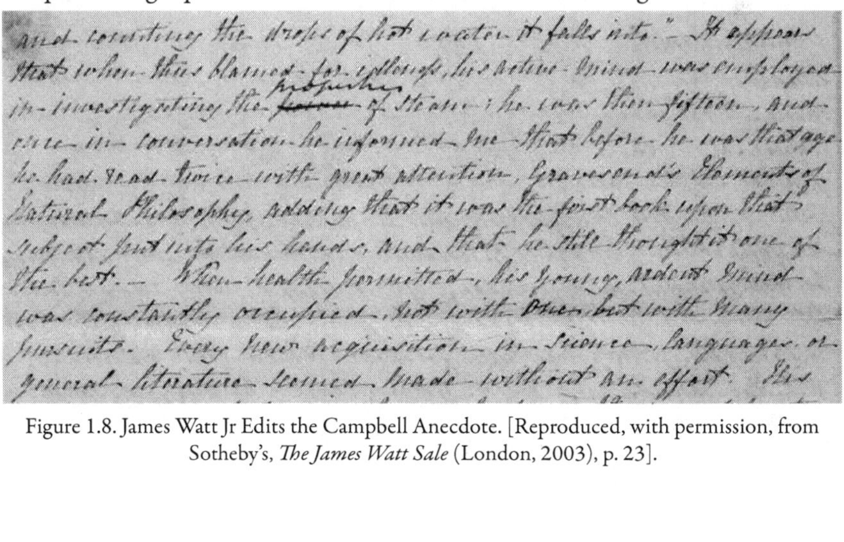
and counting the drops of hot water it falls into." – It appears that when thus blamed for idleness, his active mind was employed in investigating the ~~power~~ properties of steam; he was then fifteen, and once in conversation he informed me that before he was that age he had read twice with great attention, Gravesande's Elements of Natural Philosophy, adding that it was the first book upon that subject put into his hands, and that he still thought it one of the best. – When health permitted, his young, ardent mind was constantly occupied not with one but with many pursuits. Every new acquisition in science, language, or general literature seemed made without an effort. His

Figure 1.8. James Watt Jr Edits the Campbell Anecdote. [Reproduced, with permission, from Sotheby's, *The James Watt Sale* (London, 2003), p. 23].

mistake here. The change is identifiably in Watt Jr's hand. Watt Jr did not simply transmit the anecdote from Miss Campbell to Arago, he modified it.[39] In fact, I believe that he considered himself to be rendering it truer in so doing. Let us compare the original statement in the Memorandum with the statement as modified by Watt Jr. The original states: 'It appears that when thus blamed for idleness, his active mind was employed in investigating the power of steam'. Watt Jr's version substituted 'the properties of steam' for 'the power of steam'.

The word 'power' would have struck Watt Jr as inappropriate. After all, the power of steam had been demonstrated long before his father's day. Watt did not invent the application of steam power to practical purposes. He *improved* that application by, among other things, greatly increasing the duty of steam engines. That increased duty was due largely to his invention of the separate condenser. The separate condenser, in turn, was inspired by his father's *independent* experiments on the properties of steam. This, I suggest, was why Watt Jr made this alteration in Miss Jane Campbell's Memorandum.[40] The intent of the alteration, if not the exact words, was conveyed in the published version of Arago's *Eloge*. The exact words, as corrected by Watt Jr, were reproduced in Muirhead's *Life of James Watt*, published in 1858.

When Watt Jr made this modification of the kettle story he had already been long engaged in a fight to try to secure his father's priority over Henry Cavendish and Antoine Lavoisier as discoverer of the composition of water. Watt himself, in 1784, had expressed great resentment about the way he, and his ideas, had been treated by these two prominent natural philosophers. His son shared these concerns even as his father's anger cooled. The 'water controversy' was to explode again in a second phase in the late 1830s and 1840s as a direct result of the claims put forward on Watt's behalf, with Watt Jr's connivance, in Arago's *Eloge*.[41] It might be tempting to conclude that, in emphasizing *properties* of steam, Watt Jr was making a direct move in support of his father's claims to the discovery of the composition of water. This is all the more tempting because, as we will see, Watt's real investigations of steam were directly connected with his ideas on water. There were reasons, however, why Watt Jr did not want to make that direct connection publicly. This tactical omission is discussed further below. I believe that in altering the kettle story Watt Jr's main objective was to make a claim about his father's experimental (and therefore philosophical) approach to the improvement of the steam engine and, in particular, to assert his father's independence of Joseph Black in that regard. Of course, insofar as this established Watt's scientific credentials and capacities it did support his case in the 'water controversy' in a general sense. However, in modifying the anecdote Watt Jr almost certainly had the propagation of ideas about the relationship between the kettle and the separate condenser in mind.

The question of the scientific basis of Watt's major technical improvement of the steam engine, the separate condenser, is, of course, a classic bone of contention among historians and has involved generations of interpretation and counter-interpretation. Some historians, convinced that the technologies of the first industrial revolution were largely the product of empirics, have depicted Watt as a talented, hard-working craftsman who owed little to scientific investigations. A. R. Hall, who reserved the term 'science' for generalized, theoretical knowledge, claimed that Watt had only a veneer of science that did not meaningfully inform his works of invention. Thus Hall states:

> It is easy to overlook the fact that while particular facts and *ad hoc* correlations may be no more than awkward or intriguing to a scientist, they may be quite sufficient to permit an engineer or technologist to take an important step; conversely generalization, essential to science, may be a feeble prop for the practitioner. The literate and sophisticated engineer, like Watt, may enjoy his use of scientific labels and his mastery of scientific rationalization while always actually working with particular facts and ideas. *Any* scientific theory may envelop his actual immersion in particulars.[42]

Others, notably Musson and Robinson, have stressed the scientific nature of Watt's work, often by indicating his *reliance* upon, or at least his close association with, Black's researches on heat. Most recently Margaret Jacob and Larry Stewart have endorsed this view on broader grounds.[43] In much modern scholarship the boundaries between science/technology and between rhetoric/practice are not seen as being so readily drawn, if drawable at all. The contingent processes whereby such boundaries are constructed have become, for many, the object of study.[44]

In the later nineteenth and early twentieth century, attitudes among writers on his life to the kettle myth fluctuated with regard to the salience that they gave to Watt's scientific credentials. That is, those who promoted an image of Watt as an ingenious, hard-working craftsman, rather than as a philosopher, tended to debunk the kettle 'myth'. Thus Samuel Smiles, whilst recounting the story, disagreed with Arago's claim that it was authentic. Smiles pronounced the young Watt of the story as likely to have been 'as idle as he seems'.[45] Smiles was also influential in bringing a 'counter-iconography' into prominence. In his later years, Watt had spent many hours in a workshop that he built for himself at Heathfield Hall. Smiles gave this prominence in his accounts of Watt's life. He was effectively developing an image of Watt that projected the old craftsman of the Heathfield Hall workshop backwards, rather than the young philosopher of the kettle myth forwards, in time. A similar trend towards the craftsman view and away from the scientist or philosopher view of Watt is detectable in encyclopaedia articles on the great man in the late nineteenth century. For example, the new edition of *Chambers's Encyclopaedia* in the 1890s systematically removed material from the article on Watt that had appeared in earlier editions and had created the impression of Watt as a 'scientist'. Among the differences was that early editions contained a detailed account of Watt's steam experiments, but the

1895 edition removed this and simply stated that Watt 'hit upon the expedient of the separate condenser'.[46] As already noted the trend can be seen in the device-centred statues of Watt that became popular in the later nineteenth century. The celebrations in 1919 of the centenary of Watt's death prominently featured Watt's workshop, which was subsequently presented to the Science Museum when Heathfield Hall was demolished in 1924. There it remains as testimony to the image of Watt the craftsman. Serious scholarship backed this image of Watt and discounted the kettle myth. The foremost scholar on Watt in the early twentieth century, H. W. Dickinson, in his book *James Watt, Craftsman and Engineer*, had this to say about the kettle story:

> If an artist ever wishes to paint a genre picture of Watt, instead of perpetuating the unfounded story of his playing as a boy with the steam issuing from the spout of a kettle, he might limn the young workman in his leathern apron with the separate condenser on his lap.[47]

Whatever may have been the later cultural vicissitudes of the kettle story's plausibility, Watt himself, and his son, certainly claimed a scientific basis for Watt's improvements. They also claimed that the philosophical approach was Watt's own, independent work and not dependent, except in a general sense, on Joseph Black. What better way to do this than to put investigation of the *properties* of steam into the young Watt's mind, some years before he even met Black, so that they were there all along, merely surfacing and finding their fullest expression in the mature Watt's experiments on steam? So Watt Jr sought to constrain the interpretation of the anecdote in this way.[48]

We remain with an important question – what *kind* of investigation was Watt's inquiry into the properties of steam? The word 'properties' could refer to physical or to chemical ones. Few if any would jump to the latter conclusion in the nineteenth century because by then Watt's chemistry, at least in relation to the steam engine, was treated as an irrelevance. We can be confident that for a scientifically literate nineteenth-century audience, the precocious Watt of the kettle story was prefiguring, if anything, the adult Watt's physical investigations of steam – his insight into thermal laws and, later in the century, his prescience with regard to the laws of 'energy' and 'thermodynamics'.

Tracing the Indicator

What was the significance of the indicator held in the hand of the Watt statue unveiled at Greenock by Andrew Carnegie? (See Figure 1.9) To understand this we need to trace aspects of the history of that commemoration and also of the indicator itself in the nineteenth century.

The Greenock statue was sculpted by Henry Charles Fehr.[49] It was the second Watt statue that Fehr had produced, the first being erected in Leeds City Square

Figure 1.9. Andrew Carnegie unveils the Watt Statue at the Opening of the Watt Memorial Engineering and Navigation School, Greenock, 1 June 1908. Carnegie stands beneath the statue. Note the top-hatted gentlemen at the front of the audience and the flat-capped workers at the rear. (Reproduced from the *Greenock Telegraph*, 6 June 1908, p. 3, which reproduces it from a photograph by H. W. Jones)

in 1898. It was noted at the time of the design of the Greenock statue and building that 'A bronze statue of Watt (which, except in one small detail, is a replica of one erected at Leeds) will be erected'.[50] Inspection of the two statues shows that the 'one small detail' in which they were different could only be the replacement of the dividers held in the right hand of the Leeds statue by the indicator held in that of the 'replica'. It appears that Fehr fulfilled his Greenock commission by making this substitution.

The fact that the substitution was described by the *Greenock Telegraph* as a 'small detail', together with the failure of the same paper to remark on the device, even though it featured prominently in the drawing of the statue that it published on the day of its unveiling, might suggest that little significance was attached to the indicator.[51] Equally, it might be contended that the reproduction of the Leeds statue except in that detail is a sign that someone made a definite decision that the indicator would be particularly apposite for Greenock. This suggests, then, that the substitution was significant. To this point, however, I have found no historical records of discussions of the question between any of the key actors, namely the sculptor, the architect, the Memorial Committee, the Council and Carnegie himself.[52]

Figure 1.10. Watt Statue, Watt Memorial Engineering and Navigation School, Greenock, showing Indicator in Watt's Right Hand. (Author's photo)

Another point supporting the view that considerable thought and particular intent was involved in the inclusion of the indicator has to do with the actual design of the indicator placed in Watt's hand. It is recognizably a model dating from the very last years of the eighteenth or the early nineteenth centuries. The model depicted is what was known as a 'recording' indicator since it features the platen and weight introduced in the improved design by Watt's assistant John Southern in 1796. It is similar to an indicator in the Science Museum collection, which probably dates from between 1820 and 1840. (See Figure 6.1) All this allows us to conclude that Fehr made an attempt at historical accuracy. This suggests again that the inclusion of the indicator was not a desultory act but a calculated one.[53] Ray McKenzie reaches the same conclusion in his investigations, arguing from the statue's uniqueness:

> It is the only known monument to Watt that departs from the conventional practice of signifying his achievement through one of three common and easily recognized attributes – a pair of dividers, a steam cylinder or a governor mechanism – and thus relates directly to the business of the school to which it is attached.[54]

So, assuming that equipping Watt with an indicator was a meaningful act rather than a random one, what meanings did it convey to the top-hatted gentlemen at the front of the crowd at the unveiling and to the flat-capped workers at the rear? The history of the indicator,[55] and an examination of writings about it in the nineteenth century suggest that two sorts of conception of the device could have been present in the minds of the spectators that day and would have been presented to the students of the Watt Memorial School in its early days.

The first conception is of the indicator as a 'stethoscope'. Manuals on the indicator and its use, which were published in significant numbers in the nineteenth century,[56] devoted much attention to the indicator as a diagnostic tool. In depicting it as the stethoscope of the steam engineer they conveyed both the scientific basis of indicator usage and the element of 'art' professedly involved. Just as the experienced medical practitioner might divine much about the condition of a patient beyond the textbook accounts of symptoms, so the experienced steam engineer might see deeply into the workings of engines and be able to diagnose all manner of 'ailments' in the machinery.[57] It was thus entirely possible that the practical engineers in the crowd at the Greenock unveiling would have recognized in their memorialized, indicator-wielding townsman the sort of craftsman-engineer that Samuel Smiles had emphasized and that the early twentieth century increasingly favoured. This would have pleased Carnegie too, since he liked the notion of Watt as a practical, self-made man, and promoted that image in the biography of Watt that he had recently published. In fact Carnegie had insisted that the values of self-sufficiency and hard work be reflected in the way that funds were raised for the Memorial.[58]

On the other hand, representatives from the University of Glasgow (of whom there were a number present)[59] would have seen other significances in the memorial. Scions of the University, the home of Kelvin and other great developers of thermodynamics would associate the indicator and the indicator diagram with the scientific triumphs of the development of thermodynamics in which local scientific heroes of later generations had played such a large part. For them the indicator-wielding Watt would have represented a crucial precursor of the pinnacle of science-based engineering (and engineering-based science) of the century just past. No one would have seen the indicator as I will argue Watt originally saw it, that is, as a chemical device.

Conclusion

The overall effect of these representations of Watt in statuary and in the story of the kettle was to help to create what I call the 'mechanical Watt'. These images powerfully reinforced the textual generation of the mechanical Watt, a process that I will chart in the next two chapters, and ensured that the 'chemical Watt' remained hidden from view. The mechanical Watt came in two flavours. Some tried to construct Watt as a 'philosopher-engineer', maintaining that his work was philosophical in nature being based in experimental test and philosophical inquiry. One result of this was a tendency to find in Watt characteristics that anticipated mid-nineteenth century philosophical engineering of the sort increasingly pursued in the universities. A major difficulty with this route was finding serious philosophical activity on Watt's part other than his chemical work. Once his 'archaic' chemistry was discredited as natural philosophy there was little else by way of materials to build the philosophical Watt. Others promoted the image of the mechanical Watt as craftsman-engineer. This was an idea seized upon by Smiles, which men of the lower and middle orders of society found appealing. Proponents of both versions of the mechanical Watt were unaware of, or abandoned, his chemistry as the centrepiece of his work and marginalized it. I seek to revive the 'chemical Watt'.

There are two aspects to the textual production of the mechanical Watt – the marginalization of his chemistry, and the construction of him as an engineer. The shunting of his chemistry to one side came about partly through Watt's self-fashioning and partly as a result of the well-known 'water controversy' in which the rival claims of Watt and Henry Cavendish to the title of discoverer of that central chemical fact, the compound nature of water, were asserted and fought over. The irony is that not only Watt's opponents but also his supporters in that controversy were involved in disguising his chemical persona, one by granting too little to it, the other by granting too much.

2 THE DEMISE OF THE 'CHEMICAL WATT' IN THE NINETEENTH CENTURY

During the course of the nineteenth century Watt's chemical work and its importance to his improvements of the steam engine were effectively obscured. There was a significant disjunction between public characterizations of Watt in the late eighteenth and very early nineteenth centuries and those extant by the early twentieth. As Watt's early reputation developed and grew, contemporary sources frequently identified him as a chemist and recognized the relevance of his chemical work to the steam engine improvements. By the early twentieth century, however, Watt's chemistry had passed from view. This is readily seen in the events and publications that celebrated the one-hundredth anniversary of his death in 1919 and the two hundredth anniversary of his birth in 1936. So, by some process, during the course of the nineteenth century the recognition of the chemical basis of Watt's steam engine improvements, especially of the first phase of them involving the separate condenser, evaporated. What was that process?

A central element in this process was the 'water controversy' concerning whether Watt, Cavendish or Lavoisier should be recognized as the discoverer of the compound nature of water. I have examined this controversy, and the forces that drove it, in considerable detail elsewhere.[1] In this chapter I focus on the way in which the controversy, together with Watt's 'self-fashioning' in the later years of his life, radically transformed Watt's chemical reputation and in particular determined the extent to which he was recognized as a chemist at all.

Watt engaged in considerable recasting of his achievement in the years between about 1800 and his death in 1819. During the very early nineteenth century important investigations were being conducted which began to question the material theory of heat, and, even more acutely, the role of heat as a chemical substance. These were key ideas upon which Watt had built his original understanding of steam and of the steam engine. More particularly, from around 1790, a number of investigators, including Agustin de Bétancourt, Gaspard de Prony and John Dalton, conducted and published steam experiments, which in terms of methods, results and interpretation presented challenges to Watt's own. I believe that Watt, with aid and impetus from James Watt Jr and other members of 'team Watt', consciously sought to adjust to these developments and to reshape the image of him that passed to posterity. Part of that process was to downplay the direct role of chemistry in the steam engine story and to recast it in a form more compatible with trends that were weakening the chemical interpretation of heat. Although the theory of heat as motion rather than material substance, advocated very early in the century by the likes of Benjamin Thompson, Count Rumford and Humphry Davy, was long resisted, it did stimulate important shifts of opinion on the chemical role of heat, and especially of latent heat. The latter concept began to be dragged free of its original theoretical moorings to some extent, thanks especially to the increasing popularity of 'Irvinist' ideas,[2] and the work of the so-called 'Manchester School' around John Dalton. Watt's own narrative accounts of his steam engine improvements were the main vehicle through which the recasting took place.

One key narrative became widely available through its publication in an edition of John Robison's works in 1822.[3] Another was the so-called 'Plain Story' of Watt's improvements of the steam engine, which emerged from the archives through the writings of J. P. Muirhead in the mid-nineteenth century.[4] From that point these accounts served nineteenth-century authors, and modern historians, as apparently compelling evidence 'from the horse's mouth' of how the steam engine improvements had been accomplished. But these documents diverge in a number of important respects. Most notably Robison's account gives considerable play to Watt's steam experiments, whilst the 'Plain Story' does not. The former account found use in the characterization of Watt as a natural philosopher of the steam engine, while the latter one encouraged and supported the craftsman view of him. The reasons for these divergences become apparent when we learn the history and prehistory of the respective documents. They were the public survivors of a long line of documents produced by Watt giving accounts of his inventions. By showing that these documents were part of a conscious casting and recasting of Watt's activities, involving deliberate, but not always consistent, attempts to shape his reputation in particular circumstances, I question their value and veracity if treated as transparent historical sources.

Even as the keepers of the Watt flame downplayed the role of chemistry in the steam engine improvements themselves by, essentially, redescribing his chemistry as 'physics', or the pursuit of 'thermal laws', they maintained, indeed asserted, Watt's claim to one of the central discoveries of the Chemical Revolution – the compound nature of water. This eventually provoked the 'water controversy', which reached its climax in the 1840s and 1850s. The Watt camp was trying to claim a portion of chemical fame for the great man. However, there was no paradox involved here because the chemical fame that was sought for Watt was treated as distinct from his great 'engineering' achievement. The final outcome of the water controversy was that Watt's chemical claims became not merely distinct from his engineering achievements, but effectively extinct.

A Disjunction Demonstrated

Early characterizations of Watt as a chemist extended at least to the early phases of his monumental and textual canonization as the 'mechanical Watt'. This picture of him is a little-remarked feature of the panegyric surrounding the 1824 meeting at which the subscription was opened for the monument in Westminster Abbey. We have noted that Sir Humphry Davy in his speech to the meeting emphasized Watt's chemical credentials. Thomas Hodgskin, writing in the *Chemist*, was clear on the point:

> Mr. Watt was not only a mechanic, he was also a chemist; and we are proud of the opportunity to hold up the honours bestowed on him, as an encouragement to our youthful readers. As was happily explained at the Meeting by Sir Humphrey Davy, Mr. Watt's 'Discoveries were not owing to chance, but were founded on delicate and refined experiments connected with the discoveries as to heat, of Dr. Black. He was equally distinguished (he said) as a natural philosopher and a chemist'; and his first important discoveries were made in chemistry; and *his first important invention was an application of chemical principles to steam-engines*. His great mechanical improvements in them only began at a later period. In 1769, he took out a patent for lessening the consumption of fuel in steam-engines; and it was not till 1780 that he found the method of applying the movement of the pistons in straight lines to wheels and mill work.[5]

Hodgskin here makes explicit the claim that the Watt who took out the 1769 patent was operating as a chemist. What enabled him to reduce fuel consumption in steam engines was a clever 'application of chemical principles' based on his delicate and refined experiments.

The account of Watt in 1803 in *Public Characters*, though it did not go into much technical detail, described Watt as mechanic *and* chemist.[6] It is likely that George Chalmers was referring to Watt when, in his 1794 discussion of the role of knowledge in economic growth he symbolized the knowledge input in terms of the contributions of 'a chemist in Birmingham, a potter in Stafford, and a millwright in Manchester, who, by their genius, have *counteracted the expence and folly of the American war*'. The trilogy to which he referred, I suggest, consisted of Watt, Wedgwood and Arkwright, and in Watt's case the clear implication was that the improvement of the steam engine was attributable to the 'chemist of Birmingham'.[7]

At about the same time as Chalmers wrote this, Watt and his lawyers were preoccupied with a series of trials in which they challenged the engine 'pirates' who, they argued, were stealing a legitimately patented invention. In the course of these trials the pirates defended themselves by challenging the validity of the 1769/75 patent. In that legal context too, one line of attack by the Watt camp was to emphasize the importance of recognizing the *chemical* nature of Watt's improvements in the steam engine in his 1769 patent. When Boulton & Watt took 'engine pirates' to court for infringement of their patent, the latter defended themselves in various ways, but the chief defence involved asserting that the patent was invalid because of the insufficiency of the patent specification. A patent specification was supposed to supply enough information to enable someone 'skilled in the art' to construct the device or adopt the process in question. The claim of the defendants was that Watt's patent specification was inadequate because it did not include any mechanical drawings explaining the detailed construction of an improved steam engine. The Boulton & Watt lawyers responded by arguing that the patent was not taken out on a specific engine but on a principle of arrangement. It is important to realise that later stipulations that principles could not be patented were only in embryo in the 1790s. In fact the Boulton & Watt patent trials were to be important in the establishment of the precise nature of those stipulations.[8] In the course of developing their cases the Boulton & Watt lawyers thoroughly investigated precedents in chemical patents and sought to assimilate the Watt patent to those precedents. Mr Serjeant Adair put it this way in addressing the judge and jury in *Boulton & Watt* v. *Hornblower and Maberley* in 1796:

> I will ... recall your attention to that in which the invention did consist in order to shew you in what it does not consist. It did consist in an attention to the chemical principles and properties of steam by which the causes of the defects in Newcomen's Engine as well as the remedies were suggested...[9]

In subsequent discussions of the problematic status of Watt's patent as a patent of principle, the principle in question has been reinterpreted as simply that of arranging an engine spatially so as to have a separate condenser. This is how the principle was enshrined in drawings of the Watt engine, including the drawing we saw in the lap of Chantrey's marble statues of Watt. Because of this the origins of the idea in chemical principles have been lost sight of.

Other eighteenth-century sources that reveal the chemical Watt are the articles written on 'Steam' and the 'Steam Engine' for the *Encyclopaedia Britannica* by his old friend John Robison, who was a regular companion of Watt during the period that his improvements on the steam engine were made. Robison stated of this time: 'I had not yet studied chemistry, and Mr. Watt was my first instructor ... I was more assiduous in my attendance on Mr. Watt's occupations than ever, and studied the little model as much as he did'.[10] The latter is a reference, of course, to the model Newcomen Engine that, in the standard story, stimulated the train of experiments and reasoning that led to Watt's improvements. The implication is that in those improvements Watt was the *chemical* adept and Robison the student. The emphasis on chemistry remains throughout. When Robison referred to the numberless unsuccessful attempts to improve Watt's engine he found part of the reason for their failure to be that 'Our engineers by profession are in general miserably deficient in that accurate knowledge of mechanics and of chemistry, which is necessary for understanding this machine'.[11] Ultimately, the main way in which Robison placed Watt in a chemical tradition was by describing him as a pupil of Joseph Black working in the Blackian tradition of the chemistry of heat. We find, then, many indications in late eighteenth- and very early nineteenth-century literature that Watt's invention was widely understood and interpreted as based on a chemical understanding of steam.

By the early twentieth century, however, things were very different. We have the floor plan, and a detailed list of contents, of the exhibition held in the Birmingham Art Gallery in 1919, to commemorate the hundredth anniversary of Watt's death.[12] In it we find no sign whatsoever of Watt's chemistry. The substantial, detailed, and apparently comprehensive, commemorative volume by Dickinson and Jenkins, marking that centenary but not published until 1927, is also notable for the absence, apart from one brief reference to the composition of water, of all but bibliographical references to that aspect of Watt's work.[13] Indeed, not only Watt the chemist but even Watt the philosopher-engineer, gained little exposure in early twentieth-century celebrations. In fact the image of Watt the craftsman engineer had been so thoroughly propagated that those seeking to use commemoration to boost the funding of pure science on the basis of its ultimate utility could find little ideological value in Watt compared with a figure like Faraday.[14]

The Genesis and Function of Watt's Invention Narratives

Many histories of the steam engine and of Watt's life appeal to his own account of his steam engine improvements in the 'Plain Story' and also in his notes to Robison.[15] The great attraction of the 'Plain Story' is that it comes from Watt himself. It is usually taken as a straightforward, factual, uncomplicated telling of the real course of events. Hardly ever has the document been subjected to critical scrutiny or put into context. Remarkably, one of the few authors ever to put the 'Plain Story' in context, and cast some doubt upon its literal truth, was Robert Caird, the engineer and shipbuilder of Greenock, in his Watt Anniversary Address of 1910.[16] In fact, the 'Plain Story' was an interested document with a particular purpose. It was originally produced for use in the patent trials of the 1790s and, as Caird puts it, Watt deliberately did himself 'less than justice' because he was concerned to show that 'what he has not covered in his specifications was not patentable as being matter of common knowledge'.[17] It was also just one of a number of narratives that Watt (and Boulton) produced in the late eighteenth and early nineteenth centuries. Though Watt maintained an air of diffidence so far as fame was concerned, egged on by Boulton and later by Watt Jr, he was in fact assiduous in his attention to these matters. As Hugh Torrens has suggested, 'team Watt' could be quite ruthless and determined in shaping, even directly doctoring, the historical record in the guise of 'getting it straight'.[18]

Watt's tellings of the story of his steam engine improvements can be loosely divided into those provided before 1800, and those developed after. Prior to 1800 the stories told were closely related to the actual business of securing and defending the patents and selling steam engines. Thus, narratives were developed as part of the process of seeking from Parliament the extension of the patent. They were used in gaining publicity in Cornwall, the important early market for Boulton & Watt engines. And they were deployed in the patent trials of the 1790s, where Watt's famous 'Plain Story' originated. After Watt's retirement from the business in 1800 (and the expiry of the key patents at about the same time) the production of narratives continued unabated, but with shifting purposes. They were no longer intended to serve so directly the exigencies of business but increasingly to cement and preserve Watt's reputation. Legacy and commerce remained linked of course, but not so tightly or urgently as in the earlier period.[19]

As early as 1774–5 Watt was developing public narratives of his invention as part of the process of persuading Parliament to extend the 1769 patent.[20] This was crucial, certainly from Boulton's point of view, to the viability of the Boulton & Watt partnership. An eight-page pamphlet was produced and distributed to Members of Parliament titled 'An Account of James Watt's Improvements upon the Steam or Fire Engine'. The narrative form of this document we might call 'the dispersal of ignorance'. First it looks back upon the earliest fire engines of Savery and Worcester, and back upon the Newcomen Engine, from the vantage point of the knowledge about steam that the writer possesses. Key facts about steam are provided – its vaporous, elastic and condensable nature; the variation of its boiling point with pressure and, most sophisticatedly, the fact that 'the differences of the heats at which water boils under different degrees of pressure, increase in a less proportion than the pressures themselves; so that a double pressure requires less than a double increase of heat'.[21]

Then the operation of the Savery Engine is described, the whole prefaced by a view of the knowledge then available and the use made, or not made, of it:

> The inventors of the fire engine seem to have only known that steam was an uniform fluid, unmixed with air, and capable of being reduced to water by being brought into contact with bodies colder than itself; *and they do not seem to have attended to all the consequences of this property*.[22]

The Newcomen Engine is then depicted as a largely inadvertent solution to some of the problems with Savery's engine. In particular, the advantage that 'the cylinder is not made as cold as in Savery's construction, and therefore less steam is lost in filling it after condensation' was 'unforeseen by the inventor'.[23] The defects of the Newcomen Engine – notably the production of steam by injection-water boiling in vacuo, which steam resists the pressure of the atmosphere, and the 'destruction of steam' by the cold cylinder – were in part compensated for practically by engine operators, but not so well as they would have been if they 'had known the true principles upon which their engine ought to have been constructed'.[24]

The lack of prior experimental knowledge is presented as the key problem:

> so very few experiments had been made even upon the most essential part of the subject, that the real bulk of water, when converted to steam of a given heat, remained unknown, until the author of these new improvements determined it in the year 1764. The opinions which had been entertained concerning its bulk before that time were much beyond the truth, and could by no means be deduced from the very inaccurate experiments which were pretended to have been made. The boiling of water in an exhausted receiver at low heats was known; but it was neither known what these heats were, nor what progression they observed under various pressures, before he made his experiments upon that subject. These experiments pointed out the defects of the common steam engines, and the extent to which they were hurtful; but the means of avoiding them were not so obvious, and were the subject of much contemplation and experiment, of which these new improvements are the result.[25]

Watt's improvements, then, are depicted as being based upon his steam experiments. The claim to an extension of the patent is grounded in Watt's virtue as an experimentalist, which brought understanding and placed improvement on a philosophical basis in contrast to earlier ignorance and even fraudulence.

The form of argument that Watt used in gaining title to the improved steam engine he also deployed in efforts to sell it. Thus the account written for William Pryce, which was designed to sell the Boulton & Watt engine to Cornish mine owners, clearly drew on the earlier document. Mr Pryce, like the Members of Parliament, was treated to an account of Watt's steam experiments and their superiority to prior work in showing that 'the differences of ye heats at which water boils under different degrees of pressure, increase in a less proportion than the pressures themselves: so that a double pressure requires less than a double increase of *sensible* heat'. Further:

> The experiments which have been published concerning the bulk of water when converted into steam, are erroneous, & the opinions deduced from them make that bulk many times greater than it really is.
>
> That water boiled in an exhausted receiver at above degree of heats was known, but it was not known what these were nor what progression they followed under various pressures, untill Mr Watt determined them by a regular sett of experiments at the same time that he found the true bulk of steam ... These things being premised if we consider the common steam engine we shall find it defective.[26]

Thus the narrative form here is that experiments on the properties of steam provide a basis on which the defects of the common steam engine (the Newcomen engine) could be specified. These defects were then made good by Watt's improvements. Those improvements thus also found their rationale, indeed their cause, in Watt's steam experiments. (The perpetrator of the 'erroneous experiments' on the bulk of steam, though not named here, was Desaguliers. The critique of Desaguliers had been pursued in Cullen's lectures. As Arthur Donovan observes, that critique had more than a little about it of the chemist chiding the mechanical philosopher. To the initiated, Watt's statement here may well have been taken to echo such jurisdictional disputes, placing him firmly in the chemical camp). [27]

A year after receiving this document from Watt, William Pryce published his *Mineralogia Cornubiensis*, in which he stated that

> the only improvement which has been made in the fire engine for thirty years past, the public will justly attribute to the sagacity of Mr. Watt, whose *skill in pneumatics*, mechanics and hydraulics, is evidenced by the powerful application of elastic vapour, nearly like that of the barometer in his new constructed fire engine.[28]

Pryce remarked favourably on Watt's regular set of experiments and reported the economies claimed for the Watt engine without entirely endorsing the latter. All in all, though, Pryce's book, which provided the first public notice of the engine widely available in print, was a favourable piece of publicity for the Boulton & Watt engines at a time when their assault on the Cornish market was being hard fought. Richard Hills notes that Boulton, Watt, John Fothergill and Darwin were subscribers to Pryce's book.[29] They were supporting, and endorsing, an account of the Boulton & Watt engines that Boulton and Watt had themselves provided. We should note that when Pryce recognized Watt's achievement as drawing on 'his skill in pneumatics, mechanics and hydraulics' he was characterizing Watt's experiments on steam as pneumatic chemistry.

Further occasion for producing a narrative of Watt's steam engine improvements was provided by a rather different literary venture in 1789–90 when his friend and fellow 'Lunatic', Erasmus Darwin, requested and received from Watt, and from Boulton, an account of the steam engine improvements for use in the epic poem, *The Botanic Garden.* In his letter to Watt, Darwin asked what the world might be told about 'your improvements of the steam-engine or anything about your experiments, or calculated facts about the power of your engines, or any other ingenious stuff'.[30]

Boulton's response[31] was typically direct, giving Darwin a 'Catalogue of Facts' that he could use as he pleased, detailing how Boulton & Watt engines differed from Newcomen's. This reads very much like a marketing brochure, although Boulton also added detailed calculations supporting his contentions about the 'convenience and properties' of the improved engines. Watt's response was very different – long, diffident, almost despairing. He sent Darwin a ten-page 'dull History of Steam Engines', as he described it, and a covering letter full of signs of the ill-health and lack of 'animal spirits' that he diagnosed himself as suffering.[32] The 'dull History' was clearly composed with his earlier narratives at hand, especially the pamphlet produced for Members of Parliament in 1774. The central account of his experiments in the mid-1760s and how those clarified the failings of Newcomen engines was taken pretty directly from the earlier work. But this was prefaced by a long section on the earliest uses and understandings of steam among the ancients and in the seventeenth century. It was followed by the 'perfect engine' narrative, in which the realization of that principle was satisfied by the idea of the separate condenser.

Watt's lack of spirits at the time perhaps inclined him to a remarkably diffident passage, even for him. Watt gave Darwin permission to edit the document he sent but asked that he bestow no personal praise – 'in my own conscience I do not think I merit much praise. I am in fact a mere compilator. I have tacked and sewed together, the [serasso?] & remnants of many a mans ideas, altered & fitted them to purposes he never thought of, but I have invented or discovered little'. Interestingly, in line with this downplaying of the role of the individual, Watt included a passage at the end of his account of the steam engine narrative that strongly gestured towards deterministic ideas about invention. Referring to himself and Boulton he wrote:

> these Gentlemen owe much of their success to the accurate philosophy of the last age, to the enterprising spirit of the present age, to the opulence of this country & to the decline of prejudice & attachment to ancient [candour?]. Had they lived in another age or in some other country, they might have been mere theorists or have rubbed through life unknowing & unknown unless they had happened to turn their abilities to the contrivance of instruments for the destruction of the human species?

Even as I argue that Watt tailored his various narratives to purpose, I acknowledge that, as in this instance, their chief purpose could on occasion be therapeutic. Needless to say, Darwin knew his friend's tendency towards diffidence well and there was no sign of this determinism and self-effacement in the published account!

As we have already seen, in the 1790s, narratives of Watt's steam engine improvements were produced in connection with the patent trials of that decade in which Boulton & Watt confronted legally people whom they regarded as 'pirates' of their engines. John Robison's testimony in *Boulton & Watt* v. *Hornblower and Maberley* was his first venture into describing Watt's route to the separate condenser. In the course of his testimony Robison emphasized the philosophical nature of Watt's approach, saying that 'all became science in his hands'.

The science concerned was chemistry. This was apparent not least from the close association that Robison depicted between Watt and Joseph Black. Robison was useful in other ways when the sufficiency of the specification was being tested in the court[33] but Watt's own affadavit, which became the 'Plain Story', took the narrative in a different direction.

The story that Watt told contains much circumstantial detail but the intent of it, given the context of the trial and of earlier legal contests, was to show that virtually all the mechanical contrivances that he employed were previously known and used. The central novelty was the *principle* of the separate condenser, which in turn depended upon knowledge about the production, condensation and properties of steam. However, the 'Plain Story' played down the extent to which Watt himself had generated that knowledge. We have seen that other accounts detailed Watt's own experiments at great length, but here there is only one mention of an experiment with steam by him, the rest being garnered from others:

> By a simple experiment, Watt found what was the real bulk of water converted into steam; and from his friend Dr. Black he learned what was the heat absorbed and rendered latent by the conversion of water into steam, which the Doctor then publicly taught, and had done for some years. Experiments had been made long before by Dr. Cullen, Mr. John Robison, and others, in public classes, which proved that water, when placed in an exhausted receiver, boiled, and was converted into steam at the heat of 70° or 80° of Fahrenheit's thermometer...[34]

Thus Watt's 'Plain Story', because it was intended to support the argument that it was unnecessary and irrelevant to specify *mechanical* details in the specification, opted for a simple account of the basis of the *principle* that was the core of the invention shorn of the chemical thinking and experimentation that engendered it.[35] Most of the latter Watt portrayed as public knowledge attributable in the first place to others. He invoked the idea of the requirements of a 'perfect engine' as his basic insight.

The disjunction between the stories told by Watt and by Robison in the course of *Boulton & Watt* v. *Hornblower and Maberley* occurred despite Watt's best efforts to coach Robison about his testimony. Even as he asked Robison to appear as a witness Watt began the coaching, telling his old friend: 'I did not invent this method piece meal but all at once in a few hours in 1765 I believe ... all this was done in the day or two after I had contrived it, the other experiments you mention in the Encycl[a] were afterwards'.[36]

It appears, then, that Watt's story was 'plain' not just in the sense of being claimed 'true' but also in the sense of being simplified. When compared with Watt's earlier accounts, it downplays the extent and novelty of his steam experiments and shifts the weight towards the almost instantaneous conception of the principle of the perfect engine. To take the 'Plain Story' at face value as a simple historical description is a dangerous thing to do given that the story was constructed in a particular rhetorical context with specific legal objectives in mind. Once the patent trials were settled, the victory of Boulton & Watt's engine business was more or less complete. They dealt with the arrears owed to them, especially in Cornwall, recovering a very substantial sum. The original patent terminated in 1800. By this time, Watt Jr and Matthew Robinson Boulton were running the business affairs of a reconfigured company, Boulton, Watt & Co., the fortunes of which were now more dependent on the construction and sale of engines than on the exploitation of patents. For Watt in retirement, the focus in recounting the story of his great inventions shifted to his place in history. Reputation, rather than direct commercial issues became the prime reason for carefully scrutinizing stories told about steam. The steam experiments re-emerged to greater prominence, when compared with the way that they had been pushed into the background in the 'Plain Story'. In fact, whereas in that story Watt had deferred to Black, in post-1800 accounts Watt's independence of Black in making those experiments became a central issue once again.

Philosophical Self-Fashioning in Retirement

The publication of Joseph Black's chemical lectures in 1803, under John Robison's editorship was an early occasion prompting Watt to 'correct' the historical record. Watt was unhappy, to say the least, with the dedication of that work to him in terms describing him as 'Dr Black's most illustrious pupil', whose improvements of the steam engine 'you profess to owe to the instruction and information you received from Dr Black'.[37] Robison had made these claims in his contributions to the patent trials and, whilst generally grateful for Robison's help on that occasion, Watt did not like the way that his relations with Black had been depicted. Watt was adamant that he had not attended Black's lectures, that he had arrived at the 'facts' of latent heat independently, and only learnt about Black's prior discoveries when he presented those 'facts' to Black. The issue of the relationship between Black's work on heat and his own was an increasingly sensitive one for Watt, and not just because of the question of his philosophical independence. There must have been uncertainty in Watt's mind about the fate of his own and Black's ideas in the face of new developments in the science of heat. Rumford and Davy were early in claiming experimental proof that heat was non-material, but it was to be mid-century before they were hailed as pointing the way forward.[38] The early nineteenth-century situation was complex. Goldfarb argues that that period saw the development of intermediate theories of heat, which were neither straightforwardly material nor mechanical. Rumford's own theory was, it appears, effectively a wave theory of a sort widely discussed before the thermodynamic 'revolution'.[39]

We know, of course, that the concept of latent heat survived the vicissitudes of changing theories of heat in the nineteenth century. Black himself had been relatively non-committal on the ultimate nature of heat,[40] but Watt had become firmly committed to a material theory in which heat was a chemical substance. Sticking to these early ideas would have been increasingly difficult for Watt in the early nineteenth century. If we place ourselves in Watt's shoes in the last twenty years of his life we can begin to see how the conceptions about heat upon which he had relied in his most productive years would have appeared increasingly uncertain and, perhaps, untenable. Watt might have taken some comfort from the writings of his friend William Henry, who defended a material theory that was also a chemical one of the general sort that Watt had developed and relied upon.[41] In his *Epitome of Chemistry* Henry's prefatory panegyric on the importance of the relationship of theory and practice singled out Watt, who 'by a clear insight into the doctrine of latent heat, resulting, in a great measure, from his own acuteness and patience of investigation, and seconded by an unusual share of mechanical skill, has perhaps brought the steam engine to its acmé of perfection'.[42] But Henry had been an early convert to the new chemistry and the material of heat that he employed was Lavoisier's 'caloric'. Despite Henry's claim, the clarity of Watt's 'clear insight into the doctrine of latent heat' must have looked less certain to the man himself during this period.

Other publications gave impetus to material, but *non*-chemical, theories of heat. Notable here was the posthumous publication of William Irvine's *Essays* in 1805 by his son. Histories of chemistry routinely note Irvine's ideas as one of the lines of theory developed by Black's followers. Prior to the publication of Irvine's *Essays* his interpretation of the concept of latent heat was known mainly through his lectures – Irvine succeeded Black at the University of Glasgow – and through the published reactions of others to that interpretation.[43] Thanks to the efforts of William Irvine the younger in compiling the *Essays*, however, Irvine's material, but non-chemical, interpretation of latent heat gained greater salience in the early nineteenth century. Watt was implicated in the publication of Irvine's *Essays*. Irvine Jr relied upon, and acknowledged in the text, Watt's recollections of some of Irvine's experiments at which Watt was present as a witness. He also gave Watt credit, alongside Cavendish, for discovery of the composition of water, and he cited Watt's determination of the latent heat of steam.[44] What Irvine Jr did not do, however, which Watt must have felt mounting pressure to do, was explain how Watt's interpretation of latent heat differed from his father's.

Publication of the *Essays*, and reviews of them, prompted commentary at least on the differences between Irvine and Black. Black had interpreted 'latent heat' as the *cause* of changes of state. At the point at which water changed into steam, for example, the extra heat that had to be supplied, and that was not 'discoverable by the thermometer', was, according to Black, *combined* with the matter of water to *produce* steam. (Although Black did not, in fact, explore the broader chemical ramifications of this idea, Watt did. In that sense Watt was more heavily committed to the chemical interpretation of latent heat than the professor was). Irvine's view was that the change of state involved a change in the substance's 'capacity for heat' – steam having a greater capacity for heat than water, for example. So the latent heat was the heat that had to be supplied to 'satisfy' the steam's increased capacity for heat. The latent heat was not, on this account, combined with matter. Critics of the Irvinist view argued that it offered no explanation for the *cause* of the change of state, whilst its supporters contended that Black's account gave no reasons why the solid should have greater affinity for the matter of heat.[45] At this stage the details of these conceptual gyrations are unimportant. The key point is that the airing of these issues, looked at from Watt's point of view, must have put him 'on the spot'. Given that Watt's thinking from the 1760s to the 1790s had relied so heavily on a chemical interpretation of latent heat, and given the growing complexity (at best) and increasing tenuousness (at worst) of the place of that interpretation in chemical opinion, what was Watt to do? I suggest that he decided to simplify the situation by disengaging the story of the steam engine improvements from his original, but now increasingly problematic and apparently outmoded, chemical ideas. The 'properties' of steam was after all a highly ambiguous term. Nineteenth-century thermodynamicists often referred to the 'thermal laws' discovered by Black and by Watt as a way of putting to one side the contentiousness involved with *theories* of heat.[46] My suggestion is that Watt himself, in the early 1800s, took a step in that same direction, effectively disconnecting his chemistry from his steam experiments.

The early nineteenth century saw not only an increasingly unsettled theoretical climate in studies of heat but also some direct challenges to the way that Watt had thought of the force of steam. Watt's chemistry told him that it was the chemical combination of heat with steam that accounted for the elasticity of the latter, so that the more latent heat that steam contained the greater was its elasticity and therefore the greater the force that it would exert. But this idea began to be seriously questioned as other investigators conducted experiments on the elasticity of steam with greater precision and over a wider range of temperatures than Watt had covered in his experiments of the 1760s and 1780s. In France, Agustin de Bétancourt conducted extensive experiments, and in Britain John Dalton published on the topic in the *Manchester Memoirs* in 1802. Watt was certainly aware of this work. He had his assistant John Southern conduct further experiments on the elasticity of steam in 1796, and again in 1803. The latter experiments, and the collection together of previous ones as well as those of other researchers, may have been prompted by a request from Laplace for information on the history of Watt's invention and on his steam experiments. This request followed Watt's visit to Paris in October and early November 1802. Watt produced two replies to Laplace, which are in draft form in the archive.[47] The first, dated January 1803, is a narrative of his inventions very much in the vein subsequently published in Robison. The second, dated 22 February 1803, deals more specifically with Watt's experiments on the latent heat and elasticity of steam in response to a more specific request from Laplace. In this Watt gave an account of his earlier experiments and those of Southern whilst also displaying knowledge of the work of Bétancourt and Dalton. Watt was being put on his mettle here. It is possible that he was being tested out prior to his nomination for the prestigious corresponding membership of the French Institute, to which he was elected in 1808. Be that as it may, Watt was certainly being pressured to produce the science of steam as he understood it.

Another publication from Manchester on this question came from John Sharpe, a friend of John Dalton, a solicitor and member of the Manchester Literary and Philosophical Society, who read a paper to that body in 1806 titled 'An Account of Some Experiments to Ascertain whether the Force of Steam be in Proportion to the Generating Heat'. Although the paper was not published until 1813, it is more than likely that Watt knew of it earlier than that through his Manchester connections.[48]

Sharpe's paper presented experiments designed to ascertain the quantity of latent heat contained in steam at different temperatures. The saving of fuel in engines operating with steam at high temperatures had led him to ask whether this might be due to the constitution of steam at such high temperatures. It was known that steam at 212°F exerted a force equal to the weight of the atmosphere but that at only 40°F higher the force was doubled and that at 55°F higher still the force of steam increased to four atmospheres. Sharpe drew his figures from a paper by John Dalton.[49] So Sharpe set out to answer the following question: 'How does this comparatively small addition of temperature produce so remarkable an increase of force?' Neither published papers nor advice from friends conversant with the subject could provide an answer. However, Sharpe says:

> I was told, that Mr. Watt had made experiments; but could not learn the nature of them, or the precise results. It was supposed, however, that, as steam over water increased in temperature, it continued to combine with a greater quantity of heat than what was indicated by the mere increase of temperature; – so that steam of double the force of the atmosphere probably contained twice as much heat, in the same bulk, as steam at the common boiling temperature.[50]

Knowing that steam rising from boiling water at 212°F contains about 940°F of latent heat, it could be predicted that steam of double the force ought to contain more than 1800°F of latent heat above that indicated as its temperature, assuming that the 'stronger' steam was the same density as that produced at boiling point. So the experiments were designed to see if this was the case. If it was, then the elasticity of steam was a function of the amount of latent heat that it contained and Watt's chemical theory of heat was supported. But if there was no such relationship then the increased elasticity of steam would be better explained by an increase in its density, and Watt's position was challenged.

Cardwell sees Sharpe's paper[51] as an important, if inconclusive, examination of the source of the high fuel-economy of high-pressure engines, which it certainly is. But he misses, I think, the extent to which the paper was intended as a test of the idea that *combined* latent heat was the source of the elasticity of steam, and so a test of the chemical material theory of heat. The fact that Sharpe refers to the article 'Steam' in the *Encyclopaedia Britannica* tells us that he would be familiar at least with Robison's version of Watt's ideas on the question. Certainly, the reception of the paper on publication gives indications that this issue was the main focus of interest. For example, in its analysis of the *Manchester Memoirs* for 1813 (in which Sharpe's paper was finally published) the *Annals of Philosophy* asked, 'Now is the increased elasticity of steam owing to an increase in the latent heat of steam, or to an increase in its density? Mr. Sharpe's experiments seem conclusive in favour of the last of these opinions'.[52] In light of this sort of claim, it seems likely that Watt, in reconstructing the history of his improvements, would have been inclined to play down the chemical theory of steam that he had worked with from the 1760s to the 1790s. This was in fact what he did.

In late May 1814 Watt wrote to Dr David Brewster, who had taken over from John Playfair the task of producing an edition of the works of John Robison. Watt had agreed with Brewster to edit and annotate Robison's essays on 'Steam' and the 'Steam-Engine' from the *Encyclopaedia Britannica*, and it was this edited and annotated version that he sent to Brewster with his covering letter, itself to be published. Watt was prone to playing down his interest in this work, telling correspondents that it was a painful process.[53] Watt may well have found it a chore, but he was encouraged and helped by Watt Jr and by John Southern, who were part of a team of Watt personnel giving time to the task. From the point of view of their patriarch's reputation this was a very important venture and it was taken with commensurate seriousness. For various reasons the publication of the final four-volume work of Robison's *A System of Mechanical Philosophy* was delayed until 1822, by which time of course Watt was dead. However, some separate copies of the edited and annotated versions of Robison's essays on 'Steam' and the Steam Engine', which were the first items in volume 1 of the larger work, were printed and privately distributed in 1818. So what account did they give?

The most important general observation to make is that Watt's experiments were again in evidence because an important motive was to correct Robison's account, especially so far as Watt's relations with Black were concerned. So Watt's independent experimentation was duly canvassed. However, Watt remained cautious about stating fully the ways in which his work had gone beyond Black. The impression given is that he arrived at *the same* view of latent heat independently of the professor. Watt does not discuss the fact that he took the chemical understanding of latent heat further than Black ever did. The treatment of the understanding of steam cleverly incorporates later experimental work and makes important concessions to the Manchester School's claims whilst maintaining Watt's dignity and credibility. In this way, Watt's steam experiments were rendered reconcilable with Irvinist and Daltonian views in a way that they had never been in reality. And so, insofar as Irvinist and Daltonian views of heat ultimately proved more readily translatable into the language and conceptual world of the physics of heat and, eventually, thermodynamics,[54] so too was Watt's narrative. Let us examine the way in which this was done.

In a long note describing his experiments on the variation of the latent heat of steam with temperature, Watt acknowledged that 'Mr. Southern is inclined to conclude, from the experiments on the latent heat of steam at high temperature [presented in the Appendix] ... that the *latent heat* is a constant quantity, instead of the sum of the latent and sensible heats being so'.[55] This seems tantamount to an abandonment of what is known as 'Watt's Law' – that the sum of the latent and sensible heats is a constant. Watt had used this idea not only in his development of 'expansive working' of steam engines but also, as we will see, it was important to his ideas about the chemical transformation of water into air.[56] So, to abandon Watt's Law was to jettison a key part of Watt's chemical material theory of heat. We will see shortly, however, that a place was retained for it in a clever fashion.

A note with Southern's initials acknowledges Dalton's claim that 'the difference of the temperature of two vapours of the same elastic force, raised from two different liquids', is a constant quantity.[57] In his further account of experiments on the elasticity of steam, Watt once again acknowledged Dalton's experiments (as reported in *Manchester Memoirs*) and also those of Mr De Bétancourt (in Prony's *Nouvelle Architecture Hydraulique*), and of a Mr Schmidt. After recounting his own experiments of 1764–5 and 1774, Watt noted that he had asked Southern to repeat them, with the assistance of Mr Creighton and considered the results (presented in Southern's letter to Watt in the Appendix) as very reliable.[58]

The Appendix referred to contains Southern's letter to Watt on the steam experiments that he and William Creighton conducted in 1803. From measurements of the elasticity and density of steam, Southern stated that 'These results appear to support the conclusion that *the density of steam is nearly if not accurately, proportional to its elasticity*'. And subsequently he states: 'The opinion which I entertain from these experiments as to the *latent heat* of steam is, that it is a *constant* quantity, and perhaps this opinion obtains support from the modern discoveries of definite proportions'.[59] This is another positive gesture towards Dalton, and to this extent Southern, with Watt's agreement, is cutting ties with Watt's chemical material theory of heat. There is a caveat entered, however. Southern elaborates on what he intends 'constant quantity' to apply to and makes a distinction between what he calls the latent heat of *expansion* and the latent heat 'that may perhaps properly be called *constitutional*'. It is the latent heat of expansion that Southern considers a 'constant quantity'. He holds out the possibility that the latent heat of expansion plus the constitutional latent heat plus the sensible heat of a body of steam may be constant. This would be a modified form of 'Watt's Law', and would, Southern argued, perhaps also account for the discrepancies between Watt's measurements of elasticities at low temperatures and his own measurements at higher ones.[60] Tellingly, at this point, Watt added a note saying: 'I have, for very many years, entertained a similar hypothesis; but I know of no experiment whereby the truth of it can be demonstrated conclusively'.[61]

All this leads me to think that 'team Watt' were seeking to present Watt's experiments, and those done under his auspices by others, in a way that made concessions to more recent investigations and newer theoretical orientations, especially those of Dalton and his allies. At the same time, however, they attempted to rescue a degree of credibility for Watt's original investigations and ideas.

We have access to some of the working documents of 'team Watt'[62] and these are revealing. We can, with a little effort, distinguish Watt's original rough draft, the comments and additions of Southern, and also of Watt Jr, and the construction of the final text sent to Brewster. While I have yet to fully unravel this, there are signs suggesting the social dynamics within the team. It is evident that Southern was a 'modernizing' force, suggesting some changes to Robison's text in the way it dealt with heat that Watt did not accept. Southern was the main architect of his own letter, but throughout its negotiated composition, he was deferential to Watt, telling his master at one point that his notes were 'intended as memorandums or hints to draw Mr Watt's attention to the text, and not to guide him in the advice he may think proper to give to Doctor Brewster ... On this point Mr. Watt will of course exercise his own judgement'. Tellingly, perhaps, the draft of Southern's letter ended with the expression 'I remain, with the greatest esteem and veneration'. Before its publication Watt struck out 'veneration' and replaced it with 'respect'. Had he not done so then the independence of Southern's experiments and conclusions, and the credibility of his own endorsement of them on the question of latent heat in particular, might have been put in question.

Another author who examined the elasticity of steam and of other vapours at about this time was Andrew Ure.[63] In a paper published in the *Philosophical*

Transactions of the Royal Society in 1818, Ure gathered together the experimental results of Watt (so far as they were then available), of Dalton and of Bétancourt, as well as his own experiments.[64] Ure noted that Black was the first to investigate accurately the conversion of liquids into elastic fluids but that he 'directed his attention principally to the establishment of the general laws', leaving his pupils to investigate their 'individual applications'. Chief among those pupils were Robison and Watt, the former's investigations being made available in the article 'Steam' in the *Encyclopaedia Britannica*,

> while we have still to regret our ignorance of those executed by the latter philosopher [Watt], with probably a more complete apparatus, and more extensive views. We are indebted to him, indeed, for some curious observations of the latent heat of steam, at different temperatures, which make us lament more, the want of those on the elastic forces themselves.[65]

Ure goes on to describe his own experimental determinations of the elasticities of the vapours of various substances, and then, confining himself to the cases of steam and alcohol, he challenges Dalton's theory of the thermometric scale, concluding that 'the progressions of elasticity in vapours, taught by Mr. Dalton, are geometrical fictions, intended to quadrate with his notions concerning temperature; but not consonant with the laws or phenomena of nature?' On the contrary, Ure concluded with the view that,

> When the elastic forces of vapours are doubled, or when they sustain a double pressure, their interstices are proportionally diminished. We may consider them now as in the condition of vapours possessed of greater specific gravities. Hence, the second portion of heat introduced to give double the elastic force need not be equal to the first, in order to produce the double tension.

Moreover, this was a view, Ure asserted, that 'accords with the experiments of Mr. WATT... He found that "the latent heat of steam is less when it is produced under a greater pressure or in a more dense state; and greater when it is produced under a less pressure or in a less dense state"'.[66] Ure acknowledged that Berthollet considered this fact 'unaccountable', but it was, he contended, nevertheless fact. Whether this view of the relationship between elastic force, density and latent heat of vapours was consistent with 'chemical phenomena in general' he left others to determine. But he urged that 'Whatever be the fate of the investigation of the general law now respectfully offered, the statement of Mr. Watt may be implicitly received under the sanction of his acknowledged sagacity and candour'.[67] In affirming the value of Watt's steam experiments Ure was using the authority of the great man as a bulwark for his own experimental findings, and by treating chemical considerations as a separate issue he encouraged a view that such considerations were unconnected with Watt's experiments.

In his conclusion to the paper Ure suggested the potential value of using alcohol vapour 'for impelling the piston of a steam engine', though he prefaced the suggestion with a final tribute to Watt:

> To the theory of latent heat, which, like the hydrostatic paradox of Archimedes, might have remained for ages a barren, though beautiful proposition, the fertile genius of that philosopher [Watt] gave all at once its noblest application ... by his new steam engine. After him, many minds of the first order for science and ingenuity have offered schemes of farther improvement; but all either frivolous or abortive; with such prophetic judgment had Mr. WATT anticipated the happiest form and structure of which it was susceptible.[68]

Watt and Ure were in touch at about the time that this paper was published. Their conversation would certainly have been interesting. If Watt, as I have suggested, was faced with a problem of philosophical credibility in the light of the work of Dalton (and Sharpe, as discussed above), then in Ure he might seem to find a solution.[69] Yes, Ure defended Watt's specific findings, which would possibly have been pleasing to the great man, but more than that he affirmed a way for Watt's steam experiments to be freed of the chemical theory of heat that had given birth to them. This is even clearer from some of Ure's later writing, notably his article 'Caloric' in his *Dictionary of Chemistry*. In that article Ure drove a wedge between Watt's steam experiments and his chemical theory of heat by doing three things in juxtaposition: distancing Watt from Black on the testimony of Watt himself; supporting Watt's approach to experiments on the elasticity of steam against that of other investigators (as he had done in his 1818 paper) and, finally but crucially, taking a positivist view of the results of those experiments, declaring 'how little room there is to pronounce dogmatic decisions on the abstract nature of heat. If the essence of the cause be still involved in mystery, many of its properties and effects have been ascertained, and skilfully applied to the cultivation of science and the uses of life'.[70] It was in this sense that Ure described Watt's steam engine as 'the most splendid trophy erected to the science of caloric'.[71] For informed readers the 'science of caloric' meant the pursuit of thermal laws without too much concern about the nature of heat. They were thus encouraged to read that approach back into Watt's own work.

In this chapter so far, I have shown that there is a plausible case to be made that Watt himself wrestled with the question of how to represent his path to the separate condenser and his other improvements. He did this over many years in a variety of contexts, and those varied occasions called forth interestingly different accounts. Early versions stressed the extent and novelty of his experiments with steam, but the 'Plain Story' produced by Watt at a particular juncture in the patent trials of the 1790s took quite a different approach. His solution to the problem in the early 1800s involved him reasserting his independent experiments but now as part of a narrative that downplayed the importance of his material and chemical conception of heat and presented his steam experiments as discovering what many nineteenth-century students of heat were to call 'thermal laws'.

These moves had another important effect – they disguised the substantive connections between Watt's steam engine improvements and another chemical claim to fame as discoverer of the compound nature of water. During the course of the nineteenth century that other chemical claim was to be contested and denied, kicking further dust over the traces of James Watt, Chemist.

The 'Water Controversy' and the Demise of Watt's Chemical Fame

The de facto modernization of Watt's chemical ideas that he executed in his later accounts of his steam engine improvements also occurred in the campaign that the Watt camp mounted from the 1820s for recognition of Watt as discoverer of the compound nature of water. That campaign, and in particular the *re*contextualization of Watt's chemistry that it elicited from opponents of his claims, were eventually to prove fatal to Watt's chemical reputation. The 'archaic' chemistry thus revealed could not, of course, in all its deficiencies be granted any significant role in Watt's *real* achievements. The interpretation of those achievements was therefore driven into the arms of 'engineering' and, in some cases, 'thermodynamics', a move that the 'Plain Story' and Watt's early nineteenth-century narratives of his inventions did nothing to discourage.

In *Discovering Water* I argued that the emergence of a single individual as *the* discoverer of the compound nature of water was a cultural process of attribution that involved both an explicit public controversy and a longer, less obvious but equally important, drift of opinion detectable in textbooks and encyclopaedia treatments of the issue. Before the explicit controversy, which reached its height in the 1840s and 1850s, the drift of opinion had been towards Henry Cavendish, though Watt remained a credible candidate. The explicit controversy, which occurred mainly in the 1840s and which both camps claimed to have won, gave Watt a new lease of life in the drift of opinion, but in the later decades of the century Cavendish emerged as the discoverer in the great majority of attributions.[72]

The two camps differed crucially in their choice of ground on which to conduct the explicit controversy. Watt Jr, Francois Arago, Francis Jeffrey, Henry Brougham and J. P. Muirhead, the most active pro-Watt participants, adopted what I have called an 'empirical' set of criteria against which to judge the issue of discovery. They focused almost exclusively on the issue as one of priority – they sought to frame the controversy in terms of who first publicly advocated *the* concept of water as a compound. In stressing '*the* concept' I mean to indicate that the Watt camp believed (or at least conducted their argument as if) the chemical ideas about water that Watt and Cavendish came up with were essentially the same. They passed silently over the de facto modernization of Cavendish's ideas that had occurred in intervening decades, and they granted Watt a similar 'courtesy' of silently modernizing his ideas also. After all, their argument concerned merely the timing of the public announcement of those ideas. Certainly in Brougham's, Muirhead's and Jeffrey's cases this was a natural move for them to make given that their expertise lay in the law and not so much in chemistry. The Cavendish camp, which included most of the leadership of the British Association and a number of leading chemists, pitched their case on very different ground. According to them, evidence addressed to the issue of timing was otiose. The Cavendish camp ignored the de facto modernization of Cavendish's water chemistry, but mounted a severe critique of Watt's chemical ideas about water as 'archaic', and therefore unworthy of the status of chemical discovery no matter what might be established about the relative timing of their public release.[73]

It is worth pointing out at this juncture that while my concern in this book is to recontextualise Watt's chemistry in order to show its importance to his steam engine improvements, neither side in the water controversy had any incentive to do this and every reason to avoid it. The Watt camp followed the man himself in extricating the story of the steam engine from Watt's detailed original chemistry of latent heat. When it came to the water question they also glossed over the details of that original chemistry. The Cavendish camp were intent upon denying Watt's reputation as a chemist but were happy, even anxious, to acknowledge his engineering achievements and to have them dependent in a general sense upon 'science'. The archaic chemistry that they revealed in service of the first objective would only undermine the second if its relevance to the steam engine story was granted or explored.

The Victorian phase of the water controversy was provoked by the publication in 1839 of François Arago's *Historical Eloge* of Watt. This had been long in the making and was in many respects a collective effort by Arago, Watt Jr and Henry Brougham. Watt Jr in particular had an endless supply of energy, anecdote and evidence when it came to promoting or defending his father's reputation. Arago made a number of claims that were to anger and upset powerful constituencies in the British scientific community. He claimed priority for Watt over Henry Cavendish as discoverer of the compound nature of water. He also strongly implied that Cavendish, with the help of Charles Blagden, had purloined Watt's ideas and presented them as his own. In addition, Arago, who was a radical in politics, depicted Watt as a martyr at the hands of the British scientific hierarchy, near the apex of which sat the eccentric, frugal, but thoroughly aristocratic Cavendish.[74]

In the decades after the death of the long-serving President of the Royal Society, Sir Joseph Banks, in 1820, scientific reformers were in the ascendant, led by gentlemanly devotees of the sciences with strong links to the ancient universities. Through their institution building (notably the British Association for the Advancement of Science, f. 1831), and their writings about the state of science, its method, and its history, these gentlemen of science sought to define the nature of science and to prescribe how it should be conducted, supported and applied to the purposes of life and industry.[75] For many of these men, Henry Cavendish was a scientific hero, an icon of sophisticated, sustained, ingenious, fruitful, but thoroughly disinterested, research.[76] They were outraged by Arago's claims and accusations.

The first to respond publicly to Arago was the President of the British Association, the Reverend William Vernon Harcourt.[77] The son of the Archbishop of York, an Oxford graduate with a strong interest in chemistry, and one of the founders of the Association, Harcourt used his Presidential Address to fight back. Although Harcourt did claim experimental and theoretical priority for Cavendish, he rested his case substantially upon the superior quality of what Cavendish had done:

> Whilst the views of Cavendish are shown by the internal evidence of the experiments themselves, and the train of reasoning which they imply, to have been from the first precise, and philosophical, those of Priestley and Watt were always, as regards the former, and till after the publication of Cavendish's and Lavoisier's papers, as regards the latter, vague and wavering to a degree scarcely comprehensible to those who have not studied the ideas prevalent at that period of chemical history.[78]

Harcourt found Watt guilty of loose reasoning because his ideas on the composition of water were bound up with phlogiston theory, though he was more charitable towards Watt than other of his colleagues when he acknowledged that many of the best chemists of the day (that is of the 1780s) were similarly entangled. Crucially, however, that was not the case with Cavendish, according to Harcourt. While most around him sank in the quicksands of imponderable substances, Cavendish moved steadily 'from truth to truth, on every point on which experiments afforded ground for reasoning, unfettered by the complexity of the phlogistic theory'.[79]

Less charitable about Watt's qualities as a chemist was George Peacock, Dean of Ely,[80] who found Watt's 1784 paper in the *Philosophical Transactions* 'singularly obscure' whilst praising Cavendish's as a model of precision and intelligibility of the sort that was to be expected from someone of his powers who had been 'trained in the best and most rigorous school of inductive philosophy'.[81] Cavendish was, after all, a Cambridge man, Watt a Scottish autodidact. Another supporter of Harcourt's stance and, in fact, some thought, the secret architect of it, was the Reverend William Whewell, the Master of Trinity College, Cambridge. Whewell wrote privately to Harcourt in the midst of the furore that Harcourt's 'Address' had created – he had, after all, attacked Watt in front of the assembled worthies of Birmingham as well as the worthies of science. Whewell reassured Harcourt by denying that help was needed even as he offered it:

> I see that you are assailed on various sides for what you have said about Watt and Cavendish ... Your case, as you first put it, remains to my mind quite unshaken. But perhaps you may be wishing to know how the subject presents itself to my mind, looking at it rather with reference to the general course of the history of Science than to any special evidence of dates and the like... Watt's views are utterly damaged by involving a composition of ponderable and imponderable elements. This of itself was enough to show that he did not consider elementary composition with the rigour and distinctness which the discovery of the synthesis and analysis of bodies at that time required...

And thus Watt, who at most can only claim the merit of proposing a hypothesis, proposed one which by its very terms was unsuited to the step which science then had to take.[82]

The firmness of Whewell's rejection of Watt's credentials as a chemist is as clear from this as the rampant, purposeful presentism of his historical approach. The 'rigour and distinctness' that Whewell finds lacking in Watt was not a self-evident value in the 1780s, as he assumes it to be. Whewell's views on Watt quickly made their way into public pronouncement via his *Philosophy of the Inductive Sciences*. There Whewell first established his 'Maxim respecting Imponderable Elements', specifically that heat and light, not being things, cannot and should not be considered in chemical composition and decomposition. If they are so introduced 'vague and insecure' reasonings will lead to error. Having essentially installed the principles of Lavoisier's new chemistry as timeless ground rules, it is no surprise that Whewell finds Watt wanting: Watt 'does admit imponderable fluids as chemical elements, and thus shows a great vagueness and confusion in his idea of chemical composition'.[83]

Members of the developing chemical community took little direct part in the water controversy as such as it wound its way through the pages of the reviews. The chemists in the 1840s faced the need for a delicate balancing of the academic and practical interests in their ranks. This made full-blooded engagement with the Watt–Cavendish contest an unwise move for the leadership of the chemical community. For the most part they avoided it. Nevertheless, when the chemists founded a chemical publishing society it was called the 'Cavendish Society' and one of its first publications was a life of Henry Cavendish by the Scottish chemist George Wilson.[84]

Wilson's *Life of Cavendish* is a remarkable book for the minuteness of the detail that it enters into about the water controversy. Overall the work is a vindication of Cavendish's claims and a denial of Watt's. But in the course of its arguments the book sometimes conceded a good deal to Watt's reputation as a chemist. Wilson oscillated between the kind of rampant presentism that Whewell favoured and a thoroughgoing 'contextualism' in which he sought to apply the standards and understandings of the 1780s to judgements about the chemistry that Cavendish and Watt were producing. In his contextualist moods Wilson granted substantial credit to Watt, acknowledging that Watt's contemporaries could with justice have regarded him as a discoverer of the compound nature of water. He conceded that Watt had helped in 'inducing belief in the compound nature of water' and that his chemical contribution had 'a place in the history of science, whether it pleases us that it should have such a place or not'. But granting Watt's chemical reputation in his own time was not to deny, Wilson argued, the superior attainments of Cavendish's train of research and the superiority of his careful, judicious and accurate methods to Watt's 'following Priestley in all his devious wanderings'. It was only thanks to Cavendish's prior experiments that Watt was able to prise himself away from Priestley momentarily and 'approximate to the true theory of the composition of water'.[85]

Another writer on these questions who was sympathetic to Watt in many respects was James David Forbes, a Scottish natural philosopher known for his work on glaciation and on the science of heat. Forbes was a member of the inner circle of the British Association leadership, but not so inclined as Whewell to dismiss entirely the claims of his fellow countryman Watt. Forbes undertook to write a 'Preliminary Dissertation' for the eighth edition of the *Encyclopaedia Britannica*.[86] Forbes's theme was the interdependence of theory and practice, and Watt loomed large. We will see in the next chapter in more detail how Forbes depicted Watt as a special kind of 'philosophic engineer'. This was an important move in the construction of the 'mechanical Watt'. Our immediate concern, however, is how Forbes resolved the puzzle over what to do about Watt's chemical claims. In 1855, as he worked on the section dealing with Watt, Forbes sought and received advice from Wilson on this. The key question, as so often before in Watt's own writings about his achievements, was the relationship between Watt and Black. Wilson encouraged Forbes to reassert Watt's debt to Black so far as the chemistry of latent heat was concerned. This would set right not only writers in the Watt camp who persisted in making claims for Watt's independence of Black's work on that question but also 'Watt's own misconception of his debt to Black'.[87] This is, I think, a very significant move, since it insulates Watt's engineering achievements, which Forbes, Wilson and indeed the 'Gentlemen of Science' were anxious to praise to the skies, from *his* chemistry. Instead it made those achievements dependent upon what was by now the sanitized version of Black's latent heat discovery. The philosophic engineer drew, as was appropriate, from the university-based natural philosopher. Providing that the engineering achievements were insulated in this way, there was no harm, and much justice involved, in acknowledging Watt's chemical credentials in a general sense. Thus, while Wilson advised Forbes to leave the water controversy 'untouched in Watt's Life' he might notice with additional fullness Watt's knowledge of the chemistry of his time. He was, after all, a friend of Black and Priestley, had helped to introduce chlorine bleaching, and was an experimenter himself. For these reasons Watt 'had claims to be called a Chemist as the recognized coadjutor of the Chemists of his day'.[88] Forbes followed this advice. In this way Wilson and Forbes achieved a number of objectives in delimiting and characterizing Watt's chemical reputation: they at once granted him chemical credentials in his own time whilst denying him credit for the discovery of the composition of water (thus depriving him of a major place in the making of the New Chemistry), and severed the connection between Watt's archaic chemistry and his engineering achievements. This latter move was crucial to the viability of Forbes's attempt to construct Watt as a 'philosophic engineer'.

The supporters of Cavendish's claims in the water controversy were, in general, in a better position than the Watt camp to institutionalize their viewpoint. The explicit controversy was essentially over by the 1850s but thereafter myriads of attributions of the discovery in encyclopaedias and in textbooks built a consensus behind Cavendish.[89] Pockets of chemical recognition persisted for Watt,[90] but even at their most scholarly these had become merely an interesting sidelight on Watt's life and achievement rather than being seen at its heart. The water controversy, and Watt's self-fashioning in the last twenty years of his life, effectively extinguished Watt's reputation as a chemist and disguised the genuine links between his chemical understandings and his engineering achievements. Statues and other images of Watt, especially those drawing on the kettle story, effectively 'naturalized' the mechanical Watt and, together with accounts in popular encyclopaedias, conveyed that conception of him to a much wider audience than was directly reached by Watt's self-fashioning or by the literary slings and arrows of the water controversy. If the side of the coin bearing the 'chemical Watt' ended up face down in the nineteenth century it is obvious that the side bearing the mechanical Watt came to lie face up. I move now to examine how that face was constructed.

3 THE 'MECHANICAL WATT': THE MAKING OF A 'PHILOSOPHICAL ENGINEER'

The chemists of the nineteenth century did not embrace Watt as one of their number, but the case was very different with its engineers. In fact not only did the engineers 'pull' Watt into their company but the chemists, and other elite scientists, as we saw, 'pushed' him in that direction. Watt was, according to the likes of James David Forbes, an indifferent chemist but a 'profound' engineer. He was characterized as a 'philosophical engineer' because of his understanding, command and use of physical law.

My argument in this chapter is that Watt was made into an 'engineer' in the nineteenth century – perhaps it would be better to say 'engineers' since he was constructed as multiple manifestations of the type during that period. The implication of this process of construction is that Watt was not an engineer in the late eighteenth century, during his own lifetime. To divest Watt of the status of engineer may seem perverse and I do not mean this literally. His contemporaries often referred to Watt, and he designated himself, as an engineer. Indeed, 'James Watt, Engineer' was the extended identifier that he used in his published papers in the *Philosophical Transactions* of the Royal Society of London. His collaborative publication on pneumatic medicine with Thomas Beddoes used the same phrase on the title page.[1] But it is only apparently perverse to problematize this self-identification because the term 'engineer' underwent a remarkable evolution from the mid-eighteenth century to the end of the nineteenth. In understanding Watt's self-designation as 'engineer' we need to understand, I suggest, what that term meant in his own time. Having done that we can then comprehend something of the way that Watt habitually presented himself to the world, and how he sought to negotiate the ambiguous status that 'engineer' then brought with it. For Watt was no ordinary engineer, and he and his friends wanted to make that clear. He was a 'philosophical engineer'. Watt's knowledge of natural law and his own experiments on steam were what made Watt 'philosophical', but that knowledge was chemical knowledge to the late eighteenth century while it was glossed as knowledge of thermal laws by the early nineteenth.

The Changing Engineer

Through most of the eighteenth century the term 'engineer' was more of an occupational designation than a professional status. An engineer was in one important sense of the term an erector of engines. Watt was certainly in that business, though of course he would rightly claim to do much more. He invented and designed engines and oversaw their erection as well as making a successful business with Boulton out of their sale and operation. The only context in which the term 'engineer' rose above trade through most of the eighteenth century was a military one – 'Royal Engineers' were officers, and they were gentlemen. Only gradually and haltingly in the late eighteenth and early nineteenth century did the term 'civil engineer' come into regular use. It was intended to depict a professional man by analogy with the military engineer. The leading civil engineers claimed the gentlemanly status of their military brethren.

It is against these changing patterns of usage that Watt's self-designation as 'engineer' must be viewed. Interestingly, and importantly, he did not use this self-description consistently. The copper plate, dating from *circa* 1775, that was used to produce Watt's 'business card' was inscribed simply 'Mr. Watt, Soho, Birmingham'.[2] In the world of business, where social status counted for a great deal, for Watt to present himself as 'James Watt, Engineer' would have been unhelpful and misleading. In the learned and scientific world too, the label 'engineer' was neither widely used nor particularly advantageous to the bearer. An examination of the extent to which the term was used in nominating individuals for Fellowship of the Royal Society reveals that the *only* time it appeared before the 1820s was in the cases of 'Royal Engineers'. These men were recommended for the Fellowship not as engineers as such, but as gentlemen interested in the sciences and their promotion. A Royal Engineer was almost as likely to be admitted to the Society for his botanical, or even his antiquarian, exploits as for anything directly relating to his engineering.[3] Outside a military context the engineer was likely to be associated with trade. Under the Banksian regime in the Royal Society, and beyond, a candidate for the Fellowship who might be perceived as tainted by trade – and therefore likely to turn the FRS to illegitimate use by trading on it – had to prove in some other way a disinterested approach to natural knowledge. For the engineer this was often done by demonstrating the execution of 'public works', that is, a contribution to the common benefit or that of the State beyond personal interest or profit.[4] Watt's nomination certificate for the Royal Society does not contain the word 'engineer'. He is presented as an 'inventor' and an author:

> James Watt Esqr of Birmingham, inventor of the new Steam engine, and author of a valuable paper in the last volume of the Philosophical Transactions, being desirous of becoming a fellow of the Royal Society, we whose names are hereunto subscribed do, from our personal knowledge, recommend him, as likely to be an useful member...[5]

Yet, as mentioned, Watt did use the designation 'James Watt, Engineer' in his papers read to, and published by, the Royal Society. How can we explain this? To do so involves me in some social psychological speculation.

It is known that during his most active years Watt was given to bouts of depression ('low spirits') and to violent headaches. He lacked confidence in dealing with men, and found the rough and tumble of business life difficult and distasteful.[6] His partner Boulton, and other of Watt's close associates, accordingly diagnosed their friend as of a delicate, 'intellectual' disposition. Yet in Watt's dealings with the learned world there is a strong tone of defiance and pride. He considered that his constant appeal to experiment made him superior to other supposedly more learned predecessors, such as Desaguliers, who were less attentive in that department, and also distinguished him from mere engineers. This is evident, for example, in a draft letter that Watt intended to send to Gaspard de Prony in 1803, to give the French engineer a history of Watt's steam engine improvements. Referring to his study of the properties of steam, Watt stated: 'No man of science in this country had considered the subject since Desaguliers, & his writings on that head tended more to mislead than to Instruct. In short this very useful invention [the steam engine] was abandoned to ignorant artificers'.[7]

I suggest that when Watt, unusually and deliberately, designated himself 'Engineer' in his publications in the *Philosophical Transactions*, or elsewhere, he exhibited this same mixture of deference and defiance. He was saying, yes I am an engineer but not a mere engineer. I am superior to the 'run of the mill' engineer because I am an experimentalist and a philosopher. This was not an entirely personal act in the sense that Watt was creating space for the idea of the philosophical engineer. Yet it was, I think, deeply personal too, going to the heart of Watt's sense of identity. There are signs that Watt inscribed his books in similar fashion – for example, his copy of *Dr. Gregory's Elements of Catoptrics and Dioptrics*, an adaptation of David Gregory's University of Edinburgh Lectures published in 1715, is inscribed on the preliminary blank page 'James Watt Engineer/Birmingham/1778'.[8] Why do we inscribe books as we do? Beyond the practical business of marking property we may seek to convey an impression to the person who learns that this is our book, perhaps with one eye looking to posterity. We may even affirm our identity, in a small but significant way. Plausibly, then, in so annotating his books Watt was saying 'I am no ordinary engineer because I own and understand this work'.

If I am right in this social psychological diagnosis of Watt's decision to call himself 'James Watt, Engineer' in his publications – and it is a decision requiring some explanation[9] – then it becomes clear that in doing so Watt was making a point about the possible meanings of 'engineer' rather than describing the content of the publications to which he attached the sobriquet.

The patent trials of the 1790s were a more public forum in which Watt contributed to reshaping usage of the term 'engineer'. As already noted, the question of the adequacy of Watt's patent specification was a key issue in those trials and that issue in its turn depended upon the notion of one 'skilled in the art'. If a specification provided enough information about the invention for someone skilled in the art to be able to realize it then the specification was adequate. Boulton & Watt's legal team argued that whether someone found the Watt specification adequate or not was in fact a test of their quality as an engineer, a test of whether they were in fact truly 'skilled in the art'. Being so skilled involved, they argued, having the necessary natural philosophical knowledge and insight. The Boulton & Watt lawyers called eminent witnesses, mostly Fellows of the Royal Society, to attest to this notion of what a truly skilled engineer was. Their opponents, Hornblower and Maberley, by contrast, relied mainly on witnesses who were actual engine builders. The struggle was essentially one about what characteristics produced a superior engineer and the most credible expert witness. Was it extensive hands-on experience of the sort possessed by the engine builders, or was it a combination of such experience with a philosophical understanding of steam engines? The legal judgement favoured the latter, and in so doing publicly affirmed that Watt was a superior type of 'philosophical engineer'.[10]

So, when Watt described himself as 'engineer' in these different contexts he was making a point that an engineer could be, and in his case was, a philosopher, rather than a mere mechanic. He was claiming that an engineer could be a man of science and that the class of engineer possessing such qualities should be distinguished from the 'ignorant artificers', as Watt called the run-of-the-mill engineers. 'James Watt, Engineer' did not connote Watt's occupancy of an established role but functioned as a claim that he was pioneering a new one. The fact that Watt used this designation in his publications but not on his trade card was, I think, significant. In a business context, to call himself an 'engineer' would have both risked his being mistaken for a mere artificer and misrepresented his status as a proprietor of the firm Boulton & Watt – 'Mr. Watt, Soho, Birmingham'. It is in this sense then that I argue that Watt was not an engineer in the common eighteenth-century usage of that term. Where it might be understood in that common usage he avoided employing it. He was happy, even anxious, to attach it to himself when the term could be juxtaposed with a learned performance, or work, and so represent a claim to a newly emerging status for the engineer.

Claiming Watt for 'Engineering Science'

In the nineteenth century Watt was a potent symbol with which many sought to associate themselves and their causes. Adoptions of Watt were many, varied, and of varied importance. Some were repeated so often and sustained to such a degree that they helped to shape historical perceptions of the man and his work.

A key aspect of the professionalization of engineering in the nineteenth century was to be the claiming of learned status in the way initiated by Watt and a few other late-eighteenth-century engineers. This was an aspect, though not yet a required feature of membership, of the engineers' club founded by John Smeaton and others, in 1771. The body was called the 'Society of Civil Engineers' or, subsequently and informally, 'The Smeatonians'. Not all the members by any means were known as learned men – many of their reputations were based entirely on their major engineering works. But Smeaton himself, the figurehead, had been elected a Fellow of the Royal Society in 1753, was famous for his *Philosophical Transactions* paper of 1759 on wind and water power, was a winner of the Society's most prestigious award, the Copley Medal, in that year, and at the same time known for his major engineering works such as the Eddystone Lighthouse. Smeaton was a learned engineer. When, in the early 1770s, at a time of great uncertainty about how best to develop his engineering career, Watt pondered writing a treatise on steam experiments and the steam engine, William Small encouraged him by pointing out that 'Smeaton was first distinguished by writing'. Though in the end he did not write the treatise, Watt certainly aspired to the status of a learned engineer in the Smeatonian mould.[11]

Watt did not live long enough to see the rapid growth of professional engineering associations in the nineteenth century.[12] In fact the 'house' of Boulton & Watt as run by Watt Jr had strained relations with those associations. Privately, Watt Jr considered that Soho-trained engineers were far superior to those running the professional associations. Nevertheless, as numerous presidential and anniversary addresses testify, Watt himself became an important symbol for those organizations even though the advancement of the engineering community and profession sometimes seemed to require opposition to the idea that invention was a heroic business.[13] Watt also became an important reference point for those who promoted academic engineering or 'engineering science' in the nineteenth century. Two groups are particularly worth considering. First the 'Gentlemen of Science' – who founded the British Association and who, as we have seen, strenuously denied Watt the status of a chemical discoverer – and second the group of Northern philosophers including Rankine, Tait and William Thomson (Lord Kelvin) who played such a significant role in the development of 'engineering science' in the second half of the nineteenth century.

The 'Gentlemen of Science', such as Whewell, Harcourt, Herschel and Peacock, made their names as mathematicians, or astronomers, or as exponents of mathematical experimental philosophy, or as gentleman geologists. Thus, in general, they did not themselves pursue the practical sciences. There were important exceptions, of course. Babbage's engineering exploits in early computing are well known, and George Airy, especially as Astronomer Royal from 1836 became involved in many facets of engineering in building Greenwich into a centre for the application, as well as the pursuit, of the astronomical and geographical sciences. Whatever their own specializations, the Gentlemen of Science were all anxious to establish what they took to be the most appropriate and productive relations between the abstract and applied sciences. In this capacity, a number of them produced treatises intended to ground engineering education on a solid foundation of mechanics. In 1841, William Whewell published his text on *The Mechanics of Engineering: Intended for Use in Universities, and in Colleges of Engineers*. Although Whewell's purpose was partly to improve engineering education, he also sought to make engineering part of a liberal education, as he explained:

> If the common Problems of Engineering were to form part of our general teaching in Mechanics, this science also might become a permanent possession of liberally educated minds. Every roof, frame, bridge, oblique arch, machine, steam-engine, locomotive carriage, might be looked upon as a case to which every well-educated man ought to be able to apply definite and certain principles in order to judge of its structure and working.[14]

Whewell's point was that the custodians of mechanics were in a position to teach both engineers and the liberally educated gentleman a thing or two, and that the latter might, and ought to, be in a position to judge what many engineers did. Cambridge engineering, in which Robert Willis was another central figure, was grounded in such assumptions.[15]

George Airy made a similar point about hierarchy when, as an Honorary Member of the Institution of Civil Engineers and a member of its Premium Committee, he was involved in setting prize questions. One of Airy's suggestions for a premium topic was on steam engines.[16] He explained to his fellow committee members the background to the topic, observing that he had examined hundreds of steam engines when they were at work and often found that the engineers tending them knew little about the 'standing adjustments' of the engines: 'One of these is the state of the condensometer. Not one engine in ten has [this] in order, and not one engineer in ten has any notion of judging the state of the condensation from the temperature of the discharged water'. Airy noted that it was well known that power was lost by condensing 'too completely' but there was no rule of a theoretical or practical kind for working out the proper degree of condensation. Even when that proper degree had been worked out there might be other drawbacks in implementing it. He reported that the only reason an engineer had ever given him for not condensing more precisely was that it would 'make the engine jump'. On the basis of this experience, Airy suggested a question:

> To ascertain, from theory or from direct experiment, the degree of condensation which is most favourable for the working of a steam engine, as regards the production of mechanical power: to ascertain also whether any inconvenience is actually

> produced in any other respect by this degree of condensation, and to show how such inconvenience may be removed: and to give simple rules for the temperature of the discharged water or other indication, adapted to the use of engineers, for securing the proper degree of condensation.

Airy's purpose in this was clearly to elicit a scientific approach to the question of the ideal 'degree of condensation', but to do so in a way that recognized the practical complications and dealt with them in a scientific fashion too, and in a way that recognized the usual capabilities and levels of understanding of engineers who tended engines in order that workable rules might be formulated that they could apply successfully. This question captures well, I think, Airy's assumptions about, and aspirations for, the relationship between the scientific elite, the scientifically competent engineer and the practical working engineer. The first could set the question and judge the quality of answers to it, the second could develop solutions to it, the third could, with suitably simplified instructions, implement the fruits of the investigation without needing to understand it.

The sort of hierarchy that Airy envisaged was institutionalized in the relations and operations of the sections of the British Association that dealt respectively with the more abstract and the more practical sciences. Morrell and Thackray describe the situation thus:

> The assumption of the Association's apologists was that theory and practice were sharply distinct yet connected in certain significant ways. It was true that the mechanical arts often marched independently of mechanical science. But the best route to true, enduring, and satisfactory practice depended on theory...Theory that was encoded and manipulated via abstract symbols took precedence. Mechanical *science*, transmissible by symbolic language, and not the mechanical *arts*, transmitted by personal demonstration, was to be increasingly professed within the Association.[17]

Section G of the Association, titled 'Mechanical Sciences', was established in 1836, and appears to have operated according to this hierarchy, being more regularly led by academic engineers or people with strong scientific backgrounds. It was vital that the Association take notice of engineering since the accomplishments of the engineers were so important and visible in early Victorian Britain. That it noticed and placed engineering in a subordinate role to abstract science was a way to claim some measure of credit for the natural sciences for the technological wonders of the age and to highlight the importance of supporting ongoing investigations of nature that would, undoubtedly, produce future technological fruit.

So, where was the figure of James Watt in all this? From the perspective of the Gentlemen of Science it was important to properly appreciate the nature of Watt's achievements. The man was such a potent symbol that misunderstanding him could be very dangerous and destructive of 'right' arrangements. When the

Association met in Glasgow in 1840, the year after Harcourt had opened hostilities in the 'water controversy' in Birmingham, the question of what to say about Watt in another city where his ghost walked the streets was a vexed one. Rejecting Watt's credentials as a chemist had its dangers, especially if it was interpreted as a lack of interest in, or lack of appreciation of, his technological achievements. The General Secretaries (Roderick Murchison and Edward Sabine) instead gave the address of the President at the Glasgow meeting and invoked the 'genius' of Watt in a measured way:

> Raised through the industry and genius of her sons, to a pinnacle of commercial grandeur, well can this city estimate her obligations to science! ... she feels how much her progress depends upon an acquaintance with the true structure of the rich deposits which form her subsoil; and great as they are, she clearly sees that her manufactures may at a moment take a new flight by mechanical discoveries. For she it is, you all know, who nurtured the man whose genius has changed the tide of human interests, by calling into active energy a power which (as wielded by him), in abridging time and space, has doubled the value of human life and has established for his memory a lasting claim on the gratitude of the civilized world. The names of Watt and Glasgow are united in imperishable records![18]

At the dinner given by the Magistrates and Town Council for the Association one of the early toasts was to: 'The Memory of James Watt, and the other eminent men of Great Britain who have contributed to the Advancement of Science'.[19] Both the immediate difficulties created by the water controversy and the larger ideological issues concerning the connection of science and practice were well addressed by connecting Watt, Glasgow and Science. But just what was that connection? No one familiar with the Association's recent affairs could imagine that the science involved was chemistry! Perhaps Watt had drawn on his knowledge of science in his engineering improvements? Perhaps his development of the improved steam engine had itself encouraged scientific inquiry? That Watt the engineer be associated with science was important. If he were not then the opportunity was lost to depict one of the technological marvels of the modern world as drawing succour from science. However, what the association between Watt and science consisted in was, I think, left deliberately obscure at the Glasgow meeting. Thankfully, on other occasions and in other contexts it was thoroughly explored.

One of the central figures of the Association who devoted considerable attention to Watt and his work was James David Forbes. We can take Forbes's efforts, as reasonably representative of how the Association's leadership thought about the 'Great Steamer' and science. We saw in the last chapter how Forbes dealt with the question of Watt's chemical reputation, granting him, on George Wilson's advice, the status of a chemist as amongst his contemporaries, but not that of a chemical discoverer. I noted also that Forbes severed the links between Watt's chemistry and his engineering achievements. Here I investigate more thoroughly the structure of Forbes's treatment of Watt as a 'philosophic' engineer.

Before outlining Forbes's detailed estimation of Watt it is important to discuss the larger framework within which it was made in Forbes's 'Preliminary Dissertation' to the eighth edition of the *Encyclopaedia Britannica*.[20] As he surveyed the progress of the mathematical and physical sciences between 1775 and 1850 Forbes was struck by the 'advantages which have accrued to them both directly, and, as it were, reflexively, by the astonishing progress of the Mechanical Arts'.[21] Forbes was inclined to assimilate the mechanical arts to the physical sciences by regarding the 'masterpieces' of civil engineering – the steam engine, the locomotive, and the tubular bridge were the examples that he gave – as 'only experiments on the powers of nature on a gigantic scale'. These experiments involved inductive skills 'as truly philosophic as any effort which the man of science exerts':

> Whilst then we claim for Watt a place amongst the eminent contributors to the progress of science in the eighteenth century, we must reserve a similar one for the Stephensons and the Brunels of the present: and whilst we are proud of the changes wrought by the increase of knowledge during the last twenty-five years on the face of society, we must recollect that these very changes, and the inventions which have occasioned them, have stamped perhaps the most characteristic feature – its intense practicalness – on the science itself of the same period.

In exploring in more detail the relations of physical science and the mechanical arts, Forbes noted the great importance of instruments, of the mechanical skill and knowledge of physical principles that were united in them. The invention of instruments often involved the 'clear co-ordination of the parts of an invention towards the attainment of a given result, with a due regard to natural laws, and the properties of the substances used'. The merit of such inventions and their contribution to the progress of knowledge stood, no matter what had motivated them. Forbes noted the steam engine as 'a striking example' of this:

> Originally devised with an exclusively commercial object – the extrication of the Cornish mines from subterranean water, it became in the hands of Watt, first an instrument for experiments on the relation of heat to matter; next in its improved form a beautiful exemplification of these laws, and an enduring monument of the sagacity and skill of its author, as well as the most important inorganic agent which exists in modifying the social conditions of the entire globe. Finally ... it becomes the instrument of fresh discoveries. This very engine has a theory to be worked out, probably unimagined even by its sagacious author – its operation as an agent for obtaining power from matter by the application of heat, is shown in all probability a single case of a more general law ... and this more general theory of heat as a motive power leads, once more, to new practical deductions, to the conditions under which such machines may be most usefully constructed and employed.[22]

The elisions that Forbes makes here that deliberately or inadvertently assimilate Watt to nineteenth-century physical science are remarkable and important. Thus the phrase 'the relations of heat to matter' is used to describe the subject of Watt's experiments with the steam engine. For Watt heat *is* matter, but the phrase assumes its separate identity and easily gives the impression that there is no discontinuity involved between that fact and the beliefs of the author of the invention. It would be too big a stretch for Forbes to assume continuity with Watt when he writes about the work of Carnot, Clapeyron and Thomson, yet even here the rhetoric minimizes the gaps – consider the impact of the key phrase, '*probably* unimagined even by its sagacious author' – and encourages the reader to entertain the possibility that Watt did in fact anticipate in some fashion the laws of energy and thermodynamics.[23]

Inventors of this sort, Forbes claimed, had similar experiences as discoverers. Before he saw the 'cylinder of his model exhausted, and the piston descend under the action of his separate condenser', Watt felt the same 'throb of anxiety', and subsequently experienced the same sense of triumph, as 'the discoverer of a planet or the author of a theory'. Of course, not all inventions or inventors were like this. Mere contrivance involved no genius, nor did the actual achievement of command over nature in itself distinguish an inventor as a philosopher. What did was the 'foresight, the patience, the conceptive faculty, the clear-sighted and confident anticipations of what will be the results of natural laws acting in given circumstances, those circumstances being in essential particulars *new*'.

Within this treatment of the relationship between the physical sciences and the mechanical arts, a treatment in which Watt has already been used many times as an example, we find Forbes's characterization of the man and his work. It serves as the first topic of chapter 4, on 'The mechanics of solid and fluid bodies, civil engineering and acoustics'. Forbes began by identifying Watt as the most distinguished 'practical man of science' of the last hundred years or more. This is not so much because of the importance of his great inventions but because he 'taught men to raise the useful arts to a new dignity, – to marry them to genuine, unpretending, and inductive science, – to disparage ignorance and empiricism, and to render the labours of the workshop subservient to intellectual progress'. Watt had, thoughtfully, demonstrated the value of just that hierarchy that the Gentlemen of Science sought to institute! Before Watt the field of philosophical mechanics had been barren for many years if not centuries. Certainly there were numerous contrivances on paper but 'the really difficult theory of machinery in motion was little understood before Watt's day'. Watt was not the first to use the properties of steam as a source of power, but 'he was almost the first to study them as a philosopher, and to apply, with exemplary patience and skill, principles sought in the laboratory, to make available the most convenient of *all motive forces*'. The series of ingenious contrivances whereby he applied these principles, observing and deploying all the time the scientific relations between them, were 'a model of scientific research' as much as a triumph of the mechanical art. Forbes acknowledged that these combinations of theory and practice were not rare in the present day, but it was Watt who set the main happy example. Though Watt was by education and habit a mechanic, and also a shrewd speculator, he was unusual for someone of his profession in seeing the value of theory. This, rather than his other qualities, was the source of his success and the reason why contemporary scientific men like Forbes regarded him as exemplifying the relations of theory and practice.

In all this, Forbes does not delve at all deeply into just what the theory was that Watt brought into such productive relation to his engineering practice. Vague talk of the theory of heat, of physical principles and the like is all that his readers were offered, and this vagueness allowed for awkward issues such as Watt's commitment to a material theory of heat, let alone a chemical one, to be elided. Many of the grounds on which Watt's chemistry had been judged by Whewell, Harcourt and Forbes himself as outside the bounds of sound philosophical practice were used to rule his engineering practice as within those bounds. The mechanical Watt was awarded the title 'philosopher', which the chemical Watt had been denied.

Before leaving Forbes's treatment of Watt it is worth noticing an important exchange that he had with none other than the prominent politician and historian Thomas Babington Macaulay about the nature and the value of different types of knowledge, specifically the difference between that which was 'superficial' and that which was 'profound'. At the opening of the Edinburgh Philosophical Institution on 4 November 1846 Macaulay spoke 'In Praise of Superficial Knowledge'. He was reflecting on the value of the library of the Institution to those who were not scholars but were looking for 'useful instruction' and 'harmless amusement'. While Macaulay conceded the possibility that the library might nurture another Burns or Watt or Arkwright, he insisted that it was most likely that a broad range of the 'respectable class' would gain more modest benefits in moral and intellectual improvement. Macaulay then focused upon those who 'are haunted ... with an unreasonable fear of what they call superficial knowledge'. Such people feared that a smattering of all manner of subjects is a dangerous thing, that half knowledge was worse than ignorance. In response to this demand that only profound knowledge be valued Macaulay argued that he had never heard a satisfactory definition of profundity:

> The argument proceeds on the supposition that there is some line between profound and superficial knowledge similar to that which separates truth from falsehood. I know of no such line. When we talk of men of deep science, do we mean that they have got to the bottom or near the bottom of science? Do we mean that they know

> all that is capable of being known? Do we mean even that they know, in their own especial department, all that the smatterers of the next generation will know?[24]

Macaulay's answer to these questions was a resounding 'No!'. He argued the point that the smatterers of the present know more, and know it more securely, than even the leading men of knowledge of previous generations. Invoking the story of Gulliver, he contended that 'The pigmies of one society would have passed for giants in another'.[25] So even knowledge that is superficial by the standards of the leading intellectuals or philosophers of the time is worth having. In its turn their supposedly profound knowledge will be superseded even by the superficial knowledge of a later era.

There was little response initially to Macaulay's sally, which, though presented in a lighthearted and entertaining manner, raised highly contentious issues about the value of specialized knowledge versus that of the generalist, and the wisdom or otherwise of the widespread popularization of intellectual culture. The only immediate printed response was in *The Times* of 9 November where the leader-writer admired Macaulay's rhetoric, but concluded that the most important thing about knowledge was its mode of acquisition. It sharpened and invigorated the mind only when 'it is received with labour, directed to some practical purpose, tested with severe examination, and accompanied with some active cognate employment'. Without these conditions it produced incapable men – 'Such at least is the general complaint'.[26] A more sustained response to Macaulay came two years later in Forbes's introductory lecture to his natural philosophy class.[27]

Forbes made much of the point that superficial knowledge is sudden knowledge, that it is empirical and acquired by memory or by a limited experience. It is 'not deduced from extensive study and general principles'. The first discoverer of something always has to work harder than the improver and is to be admired more, Forbes contended. It was Watt who sprung to mind as an example: 'the leaky model of the steam engine on which Watt first exercised his skill, is a greater trophy of power than all the machinery in Soho'. As we saw above, Forbes was later to argue in his 'Preliminary Dissertation' to the *Encyclopaedia Britannica* that it was sustained application of physical principles to the solution of a practical problem that constituted Watt's genius. The result was profound knowledge.

The bearer of superficial knowledge might amuse us and gain our admiration in much the way that a juggler or conjurer does, 'but we should not employ him to build our ships, or to direct our railways. We engage an engineer of education and ability who is perhaps wholly devoid of manual dexterity himself'.[28] Though Watt combined qualities of head and hand in unusual measure, it was the head that made the difference. Forbes urged his students not just to acquire facts but to seek a deeper understanding of principles in order to avoid superficiality in their knowledge.

Forbes's riposte received some attention, including an anonymous review in *Fraser's Magazine* that was in fact by his friend Whewell.[29] The Master of Trinity did not produce his best performance here but he did emphasize that Macaulay's argument about the relativity of what could be judged superficial was delusive because it involved treating knowledge as a commodity, as bricks in a wall. On the contrary, Whewell argued, knowledge grows not by additions but by transformations. Profound knowledge – a term that Whewell did not entirely like – involved 'distinct and clear thoughts, not merely remembered words; knowledge connected with principles, not merely noted as facts'.[30] These were the criteria by which Whewell, like Forbes, judged Watt's chemistry as superficial but his engineering profound, though it is ironic that, in arguing against Macaulay, Whewell granted an integrity to past knowledge systems that should have made him more sympathetic to the chemistry of imponderables.

Whewell and Forbes were in fact not entirely in accord on the question of superficial versus profound knowledge, as their private correspondence reveals. Whewell tried to dissuade Forbes from publishing his Introductory Lecture, perhaps thinking it unwise to tangle with Macaulay.[31] Whewell thought that they disagreed on comparing the knowledge of earlier and later times but Forbes insisted that 'my thesis is exactly that which you so well state that "the knowledge of later times is more advanced, more extended: but not more profound, necessarily, except in more profound minds"'.[32]

Let us turn now to the group of Northern philosophers who promoted their own brand of 'engineering science' from the middle decades of the nineteenth century. The group have been thoroughly characterized by Smith and Wise as activated in their development of energy physics by theologically based ideas about waste and economy.[33] There were important differences between the Northern philosophers and the Cambridge group of Gentlemen of Science. While the latter bequeathed strong suspicions about mixing commercial and scientific objectives that were taken up by the later nineteenth-century advocates of 'pure science' such as Huxley and Tyndall, the Northern philosophers had no such reservations. William Thomson (Lord Kelvin from 1892) personified the amalgamation of roles as natural philosopher, engineer and businessman. It is revealing that Forbes decided to decline an invitation from Thomson to attend a public dinner in Glasgow in connection with the Atlantic Telegraph, explaining to Whewell: 'That, as you know is not in my line. And while cordially rejoicing in our excellent friend's distinction, I cannot help severely regretting that he has mixed himself up so much with the commercial branch of the affair – which besides is far from prosperous. He is not only a Director but a Patentee in connection with it'.[34] Macquorn Rankine was another of the group who sat astride the worlds of science, engineering and business. Having won prizes in Forbes's natural philosophy classes at the University of Edinburgh in the mid-1830s, Rankine went on not to the Cambridge mathematics tripos but to an engineering apprenticeship with John Benjamin MacNeill. Thereafter he combined a varied engineering career with high-level science publication. He read Clapeyron in 1842 and then began his writings on elasticity and the mechanical action of heat, culminating in the early 1850s with his thermodynamic formulations. These were notable for linking the new science to the thermodynamics of engines in a way that gave, simultaneously, theoretical insight and practical methods of determining engine efficiency. It was Rankine too, who developed the 'diagram of energy', which used the theory of the indicator diagram to unite the scientific study of energy and the practical operation of heat engines. This was the brand of 'engineering science' that Rankine promoted when he succeeded Lewis Gordon as professor of civil engineering and mechanics at Glasgow University in 1855. In that position he advanced the cause of 'engineering science' through his classes, his texts, his work for the British Association, the Institution of Engineers in Scotland, and the Glasgow Philosophical Society. He did much work on naval architecture and hydrodynamics, which placed him at the centre of a thriving academic engineering and shipbuilding community in Glasgow, a community much given to the use of Watt as a cultural symbol in the late nineteenth century.[35]

The ideology of the relations of theory and practice espoused by the Northern philosophers was interestingly different from that of the Gentlemen of Science. Whereas the latter tended to treat theory and practice as distinct but related realms, the former emphasized their unity. Consider, for example, Rankine's introductory lecture on the 'harmony of theory and practice in mechanics', which was published in 1856 on his accession to the Glasgow chair.[36] In that lecture Rankine traced ideas about the relations of theory and practice since the time when the Ancients put them in opposition, committing the great fallacy of a 'double system of natural laws', the one type being theoretical, celestial, rational, and the other practical, terrestrial, empirical. From the time of Archimedes, he argued, the idea long persisted that mechanical skill and practical knowledge were of an inferior order, and that scientific men were 'unfit for the business of life'. Rankine conceded that it was hard to find in his own day anyone who would rationally defend the 'discrepancy of theory and practice' but he argued that people continued to act as if that discrepancy was real. As an example, he offered a quotation from Macaulay,[37] (as seemed to be *de rigueur* in introductory lectures at Scottish universities at the time!) which *contrasted* the deployment of geometrical skill with practical accomplishment in just that way. Rankine's response is worth quoting at length:

> In fact the mathematical theory of a machine, – that is, the body of principles which enables the engineer to compute the arrangement and dimensions of the parts of a machine intended to perform given operations, – is divided by mathematicians, for the sake of convenience of investigation, into two parts. The part first treated of, as being the more simple, relates to the motions and mutual actions of the solid pieces of a machine, and the forces exerted by and upon them, each continuous solid piece being treated as a whole, and of sensibly invariable figure. The second and more intricate part relates to the actions of the forces tending to break or to alter the figure of each such solid piece, and the dimensions and form to be given to it in order to enable it to resist those forces: this part of the theory depends, as much as the first part on the general laws of mechanics; and it is, as truly as the first part, a subject for the reasonings of the mathematician, and equally requisite for the completeness of the mathematical treatise which the engineer is supposed to consult. It is true, that should the engineer implicitly trust to a pretended mathematician, or an incomplete treatise, his apparatus would come down in ruin, as the historian has stated: it is true also that the same result would follow, if the engineer were one who had not qualified himself, by experience and observation, to distinguish between good and bad materials and workmanship; but the passage I have quoted conveys an idea different from these; for it proceeds on the erroneous supposition, that the first part of the theory of a machine is the whole theory, and is at variance with something else which is independent of mathematics, and which constitutes, or is the foundation of, practical mechanics.[38]

This fallacy Rankine regarded as being at the bottom of a number of evils in modern society especially the lack of mutual communication between men of science and men of practice. He lamented both the tendency of men of science to favour and relish the abstruse and of practical men to over-value the practical skill acquired by observation and experience in business by rejecting what science could contribute. But on a hopeful note he considered that this lack of communication was in decline. Evidence for this, Rankine thought, lay in the founding of university chairs in engineering science and also in the mutual communication between the Mechanical Sciences and Physical Sciences sections of the British Association for the Advancement of Science. The Northern philosophers became a powerful force in the Association from the 1850s.

In the passage in his *History of England* that Rankine objected to, Macaulay had been making a point about the 'science of Politics':

> What the engineer is to the mathematician, the active statesman is to the contemplative statesman. It is indeed most important that legislators and administrators should be versed in the philosophy of government, as it is most important that the architect, who has to fix an obelisk on its pedestal, or to hang a tubular bridge over an estuary, should be versed in the philosophy of equilibrium and motion. But, as he who has actually to build must bear in mind many things never noticed by D'Alembert and Euler, so must he who has actually to govern be perpetually guided by considerations to which no allusion can be found in the writings of Adam Smith or Jeremy Bentham.[39]

Rankine rejected the view of engineering that Macaulay used in this analogy. For Rankine, the 'engineering science' that he pursued came from, and considered, all that was practical and so could return its theory to the practical without fear of omission or of the need for an addition of what we would call 'tacit' knowledge.

The theory and practice of the steam indicator was an excellent example of Rankine's point, and his treatment of it is an exercise in trying to render the tacit dimensions of indicator use explicit and to analyse them scientifically. It is significant that Rankine did not employ, or countenance, the notion of the indicator as stethoscope, a notion that was very popular with other writers on indicators in the second half of the nineteenth century. The analogy with the stethoscope conceded too much to an art of engineering reliant on the tacit knowledge of experience. I would agree that Harry Collins's treatment of such matters suggests that Rankine was doomed to fail, but for the present point it is the fact that he tried to eliminate the tacit that is important. Engineering science was not distinct from, or above, practical engineering, it simply *was* practical engineering of a new and superior kind.[40]

Where does Watt come into this? Well, his example provides the capstone for Rankine's case:

> For a century there has stood, in a room in this College, a small, rude, and plain model, of appearance so uncouth, that when an artist lately introduced its likeness into a historical painting, those who saw the likeness, and knew nothing of the original, wondered what the artist meant by painting an object so unattractive.
>
> But the artist was right; for ninety-one years ago a man took that model, applied to it his knowledge of natural laws, and made it into the first of those steam-engines that now cover the land and the sea; and ever since, in Reason's eye, that small and uncouth mass of wood and metal shines with imperishable beauty, as the earliest embodiment of the genius of James Watt.
>
> Thus it is that the commonest objects are by science rendered precious; and in like manner the engineer or the mechanic, who plans and works with understanding of the natural laws that regulate the results of his operations, rises to the dignity of a Sage.[41]

Thus Rankine, like Forbes, presented Watt's experiments with the model steam engine given to him by Professor Anderson to repair, as a model of the relations of theory and practice. For Rankine, Watt became the model 'engineering scientist'. This was Rankine's version of what profundity in engineering consisted in.

In the slightly fuller treatment of the history of Watt's improvements which he gave in his *Manual of the Steam Engine*, Rankine again gave prominence to the model Newcomen engine and described Watt's procedure as follows:

> Watt set to work scientifically from the first. He studied the laws of the pressure of elastic fluids, and of the evaporating action of heat, so far as they were known in his

> time; he ascertained as accurately as he could, with the means of experimenting at his disposal, the expenditure of fuel in evaporating a given quantity of water, and the relations between the temperature, pressure and volume of the steam. Then reasoning from the data which he had thus obtained, he framed a body of principles expressing the conditions of the efficient and economic working of the steam engine, which are embodied in ... the specification of his patent of 1769...[42]

So Rankine has Watt moving smoothly from the puzzle of the model Newcomen engine to experiments, to data, to principles, and thence to specification of the improved engine. Once again history is cast in the mould of the ideals of engineering science practice.

Rankine's successors continued the tradition of deifying Watt. Alexander Barr, who was to be the University of Glasgow's official representative at the opening of the Watt Memorial Engineering and Navigation School, was one of these. In 1889 Barr had been appointed to the professorship of civil engineering and mechanics, which Rankine had formerly held. As in Rankine's case, so in Barr's, Watt loomed large in his inaugural address, which offered a retrospective on the mechanical arts. In such a retrospective and account of the aid which science had given those arts, Barr said, 'we cannot fail to be led back to the genius of James Watt as the fountain-head in which the stream of progress had its source. With Watt, Engineering, as we know it, may be said to have begun'.[43] Barr found in Watt's forebears qualities which had led to the blending of theory and practice 'in the very flesh and blood' of the boy Watt. Watt's acquisition of knowledge of natural philosophy – through his own study and in learning from his contacts at the University – was crucial to his approach. Invention, as in Newcomen's case could be the result of accidental observation, but in Watt's case 'the man of science, made his great invention by reasoning founded upon the knowledge of the laws of nature, which he had acquired through a long course of study and experiment. There was no accident here, his method was bound to lead him to the desired result'.[44] The university had played its part in this, and, more recently, in supporting the work of Sir William Thomson, whose researches were carried out 'in the very rooms in which Watt worked and studied'. Barr considered that the founding of schools of engineering science around the country was testimony to the value that engineers now placed 'upon such a course of study as that which enabled Watt to strike out into new and untrodden paths in invention and discovery'.[45]

Watt could serve other objectives besides those of the promotion of engineering science. John Scott Russell, the engineer, naval architect and collaborator with Isambard Kingdom Brunel on the *Great Eastern* steamship, gave one of the early Watt Anniversary lectures to the Greenock Philosophical Society in 1867.[46] Russell, who had then-current controversy about the patent system on his mind, took a rather different and unusual angle in depicting Watt as a 'philosophical engineer'. He relied on a distinction between discovery and invention

– the latter involving merely a 'cunning piece of ingenious mechanism' (what Forbes had called 'contrivance') for which a patent might be given, the former a 'hidden truth' which, being a principle, 'all men may, and do, and must use it'.[47] Of Watt's status Russell was certain:

> James Watt was a discoverer. He discovered the great principle of condensation – *instantaneous isolated condensation*. He discovered that *vacuum* was a commodity that could be housed, garnered, stored, husbanded, set apart, kept at hand ready for instantaneous use ... [This discovery] was enough to have made him celebrated as a philosopher, even if he had been ignorant as an inventor, and incapable as a mechanic.[48]

Russell justified this 'singular' view of Watt's achievement by using once again the familiar stories of the repair of the model engine at the University of Glasgow, the ensuing experiments with steam, the conceptualization of keeping the cylinder always hot whilst still (and contrarily) producing a good vacuum, and the moment of inspiration on Glasgow Green when Watt supposedly had the idea of the separate condenser that would achieve this. Russell thus relied primarily on Watt's 'Plain Story'. Nevertheless, pinning down just what Watt's discovery consisted in involved Russell in some difficulties. At one point he identified it as the 'great discovery of the annihilation of steam in a separate vessel, so as to make it a never-failing, never-ending reservoir of vacuum'. At another Russell noted that 'What he [Watt] claims to have discovered is the true nature of steam', by which Russell meant 'properties' of steam that dictate the most effective arrangement of the parts of the engine. If the discovery is of the properties of steam then it would seem that the spatial arrangement of the parts of the engine is an invention. However, Russell insisted on calling this spatial arrangement – the 'isolated condensation' – a discovery also, because he wanted to place it beyond the reach of patents. He was writing at a time of considerable controversy over the patent system, and his remarks on patents should be seen in that context. Russell distanced Watt's achievement from the patent system:

> his great difficulty in obtaining a patent, and in maintaining it against evasion, consisted in this, that he had discovered a principle, that no principle can be patented, and that any patent for any machine he might invent could be so easily evaded that his right to one really turned on the delicate legal quibble whether it was possible apparently to take out a patent for a contrivance, whereas in reality it was a discovery he had made and not merely a machine.[49]

So here the symbolism of Watt the 'philosopher engineer', the discoverer *above* the inventor and mechanic, was used by Russell to try to undermine the patent system and argue for its abolition. Such arguments were to be common among patent abolitionists of the period.[50] Russell's treatment of Watt illustrates the flexibility of Watt's career and persona as an ideological resource. Though he could be mobilized in this way by patent abolitionists like Russell, he was more often recruited to the pro-patent cause. Those who sought to reform the patent system by strengthening the position of the patentee used Watt's precarious patent career as an object lesson in the need for increased protection if crucial inventions were not to be lost entirely.[51]

My own argument would place discovery truly with the nature of steam. For Watt the properties of steam follow from its nature as a chemical compound of heat and water. That idea, though, can find no place in Russell's account of Watt's discovery. His Watt, and that of many others among the elite of nineteenth-century engineers, is not a chemist, but an experimentalist who studied thermal laws, an exponent of 'engineering science'. I will examine in Chapter 6 how a number of engineers and thermodynamicists depicted Watt's ideas about heat. For the moment let me discuss some more examples of the placing of Watt so far as the relations of theory and practice are concerned. An example from the later nineteenth century is provided by the case of Sir Frederick Bramwell, who wrote the entry on Watt for the *Dictionary of National Biography*.

Bramwell was an engineer of a different stamp from the academically based but commercially active Northern philosopher-engineers. The son of a London banker, Bramwell was apprenticed to the mechanical engineer John Hague, working primarily on rail projects but also on experimental atmospheric railways and on an abortive scheme for steam-propelled road vehicles. He then managed an engineering factory before setting up his own business as a consulting engineer in the early 1850s. He became renowned as an expert witness in legal, especially patent, cases, his success being due to a knack for straightforward explanation of complex technical details and a good line in engaging exposition. Parliamentary Committees were another of his regular haunts and he represented numerous companies engaged in civil engineering projects. Bramwell was also very active in engineering and scientific institutions, including the Institutions of Mechanical Engineers, Civil Engineers and Naval Architects, the Society of Arts and the British Association, where he was twice chairman of Section G and, in 1888, one of the few engineers to preside over the entire Association.[52] When Bramwell was elected a Fellow of the Royal Society in 1873 he would have been already acutely aware of the tensions in that organization between the advocates of 'pure science' and the exponents of 'engineering science'. His comments on the relations of science and practice in public addresses reflected his professional trajectory, which placed him as an outsider to both camps. He was inclined to see the inventor as *sui generis*, telling the Society of Arts in 1874, for example, that inventions, as distinct from scientific principles behind them, were most often the product of the 'ingenious' man, who, in the course of his regular work conceived an improvement. Bramwell contended that 'Between the discovery and its application there is a gap over which none but the workman can throw a bridge'.[53]

Bramwell's Presidential Address to the BAAS at its Bath meeting in 1888 was an eccentric, rambling performance but nevertheless had a serious message to convey about the relations of science and engineering. In what was fairly clearly a jibe at T. H. Huxley and his associates, Bramwell set out expressly to counter the sentiment of the 'Pure Scientist':

> I trust ... I shall prove to you that applications of Science, and discoveries in pure Science, act and re-act the one upon the other. I hope in this to carry the bulk of my audience with me, although there are some, I know, whose feelings, from a false notion of respect for Science, would probably find vent in the 'toast' which one has heard in another place – this 'toast' being attributed to the Pure Scientist – 'Here's to the latest scientific discovery: may it never do any good to anybody!'[54]

Bramwell explained that he, an engineer, came to address the Association as President by virtue of a system that informally and approximately rotated the office between its Sections. It had been decided that it was the turn of Section G, Mechanical Science. Bramwell noted that the last time a 'purely Engineering' member occupied the chair was in Bristol in 1875 when Sir John Hawkshaw was President. He was disinclined to count Sir William Siemens's Presidency of the Association as an instance since that great engineer's scientific attainments saw him representing the Physical Section A, rather than Section G. This was symptomatic, Bramwell argued, of a larger problem:

> when Engineers, such as Siemens, Rankine, Sir William Thomson, Fairbairn or Armstrong, make a scientific discovery, Section A says it is made, not in the capacity of an Engineer, and, therefore, does not appertain to Section G, but in the capacity of a Physicist, and therefore appertains to Section A – an illustration of the danger of a man filling two positions of which the composite Prince-Bishop is the well-known type.[55]

A greater willingness to contemplate the engineer as part of the scientific enterprise required, Bramwell argued, a better appreciation of how the advancement of science and its application were related. Section G, he contended, contributed more than any other to the application of science and, for that reason, 'is doing a very large share of the work of *advancing* Science, and is fully entitled to be periodically represented in the Presidency of the whole Association'.[56]

Bramwell used substantive developments to support his case. No one, he suggested, could argue that the commercial application of electricity to communication and to the purposes of lighting had not given an 'enormous impulse' to 'pure electrical science'. The possibilities and probabilities of applications had stimulated acute minds to investigations in electrical science. He also discussed the development of prime-movers, which were due to the application of science and the work of the civil engineer. (Bramwell noted that he used that term, as it should be used, to apply to *all* engineering other than military.) In fact, he recalled, the Charter of the Institution of Civil Engineers defines civil engineering as 'the art of directing the great sources of power in Nature for the use and convenience of man'. Nothing could be more beneficent or honourable despite the tendency of many to treat such work as mechanical and unintellectual. Bramwell struck back against the purists. Some would make the engineer subordinate to even the mere bookworm, but Bramwell reminded them of the words of Sir Thomas Browne in condemning those who made their 'head a grave'. Far from this, the engineer must not only have, but must exercise science, and he must be trained in it to appreciate what might be done with scientific discoveries. In this way 'the application of science by the engineer not only stimulates those who pursue science, but adds him to their number'. Bramwell even argued for the poetry in engineering, what today we might call the 'technological sublime',[57] and concluded:

> the work of the Civil Engineer is not of the earth earthy, is not mechanical to the exclusion of science, in not unintellectual; but is of a most beneficent nature, is consistent with true poetical feeling, and is worthy of the highest order of intellect.[58]

All this might make us curious as to how Bramwell chose to portray James Watt in the *Dictionary of National Biography* article that he was asked to write.[59] Bramwell, in fact, made Watt something of a composite of otherwise competing depictions. First, at the head of the article, he designated Watt unequivocally, and uniquely, as 'engineer' in a manner reminiscent of his own assumed status in the Association as a 'purely engineering' member. Bramwell also made a good deal of Watt's 'manual dexterity' as exhibited in making models in his father's business. He also noted that Watt 'prepared himself, as a man of science, to deal with the problem of improvements of the steam engine in actual practice'. This suggests that Watt conformed to Bramwell's stated ideal for the engineer to be a generalist rather than a specialist so far as science was concerned. Indeed, in summarizing Watt's character as an engineer he offered the view that he was 'not a mere specialist dedicated to one subject, but was of great general scientific learning, and was a happy instance of a man who based his inventions on scientific data, and proved them in the model form by aid of his rare manual dexterity'.[60] Understandably, given his own origins, training and mature professional persona, Bramwell opted for a composite Watt, an engineer who drew upon science but also an ingenious man who had the qualities of the craftsman.

At the very apex of engineering science in the later nineteenth century stood Sir William Thomson, the most prominent of the Northern philosopher engineers, who became Baron Kelvin of Largs in the New Year's Honours List in 1892. Congratulations flooded in from all and sundry. Fellows of the Royal

Society, among them T. H. Huxley, slapped their colleague on the back and revelled in what they described as the honour done to science. W. H. Preece, electrician to the Post Office and about to become its engineer-in-chief, addressed his letter of 'sincere pleasure' at Thomson's 'translation to the upper sphere' to 'Baron Megohm & Lord of the Lightning'. He considered that translation to be an 'honor which the whole of our profession [engineering] will regard with gratitude for it crystallizes imperically [*sic*] what we all feel mentally that you are not only our respected chief but our much loved leader'. Members of the naval service welcomed this reward 'for your splendid compass to steer us clear of danger, and your invaluable sounding machine to guard us'. Others considered Thomson deserving for his 'extraordinary inventive genius and profound scientific research'. A correspondent who had studied with Thomson's father observed that Arago had spoken in his life of Watt with 'indignant surprise that Watt had not been made a Peer', a reproach that had been taken away with Thomson's own elevation.[61] The new Baron Kelvin of Largs shared the protean reputation of the 'Great Steamer'.

Kelvin, it might be thought, had little need for symbolism at any public event; he had merely to show up and be himself. But to have Kelvin and Watt on the menu together was a heady combination, which proved tempting a number of times as the century turned. *The Scotsman* reported that on 22 January 1898 the fiftieth Watt Anniversary dinner had been held in Glasgow at the Windsor Hotel in St Vincent Street with a distinguished company of seventy gentlemen drawn variously from the British and Japanese navies, the engineering and shipbuilding worlds of Glasgow, and headed by the Provost and Lord Kelvin himself.[62] Kelvin was called upon to propose the toast of the evening, 'To the Memory of James Watt'. Here is part of the newspaper's report:

> There was in the possession of the Glasgow University still, he [Kelvin] said, the model which Watt was employed by it to repair, and in repairing which the great man saw not only what was to be mended, but some of the defects of the principles which it realized, and was led from that small beginning to the great work which they were that evening commemorating – (applause) – a work which had made the century, not in the eyes of engineers merely, but in the eyes of the whole world, the century of James Watt (Applause) ...Of all the sixty centuries of recorded human history no century had been so great as the 19th century in engineering, in the work of the engineer to apply the laws of matter and the principles of science for the benefit of mankind. (Applause) ... In pure science Watt was a leader. He (Lord Kelvin) well remembered how the words 'La loi de Watt' were always on the lips of his (Lord Kelvin's) great master in physical science, Regnault. They referred to a decision of Watt's in regard to the latent power of steam, as to which different opinions had been given. Watt showed himself a man of penetrating genius in many branches of science...[63]

In this way, Kelvin publicly celebrated Watt as, like Regnault, a great master of physical science, which quality rendered him, as an engineer, a great benefactor to humanity. It is worth observing, incidentally, in relation to the themes of Chapter 2 above, that Victor Regnault's endorsement of 'Watt's Law' was, given Regnault's famously definitive work on steam, an important instance of the modernization of Watt's original, chemically-based steam experiments that obscured the original meaning of that Law.

In May 1901 we find Kelvin reading Muirhead's *Life of James Watt* with much interest preparatory to giving an oration on Watt at the Glasgow University Jubilee celebrations the following month. Kelvin's letterbook of the period contains his draft of the oration, which was delivered on 12 June 1901.[64] Once again the *Scotsman* reported fully on the event. While Kelvin's draft oration, and its published version,[65] had a great deal to say about the origins and manifestations of Watt's scientific approach to steam engineering, telling a fairly comprehensive tale up to Watt's departure for England, the oration as delivered may have been somewhat briefer. Certainly, the *Scotsman* stressed those sections of it which displayed Kelvin's claims about Watt's prescience in relation to recent developments. Thus it reported that, having mentioned Watt's experiments with a Papin's digester and a syringe in the early 1760s, Kelvin remarked that that simple arrangement had formed the basis of engines ever since:

> Precisely that single-acting, high-pressure, syringe engine, made and experimented on by James Watt one hundred and forty years ago in his Glasgow College workshop, now in 1901, with the addition of a surface- condenser cooled by air to receive the waste steam, and a pump to return the water thence to the boiler, constituted the common road motor, which, in the opinion of many good judges, was the most successful of all the different forms which had been made and tried within the last few years.[66]

Thus, Kelvin held up Watt as responsible not only for the improved steam engine of the nineteenth century but also for the anticipated steam motor vehicle of the twentieth! When Kelvin spoke, the battle between steam, electric and internal-combustion engines was in fact still being fought and his projection of steam as potential victor in that contest, unlike his award of credit for it to the Watt of the 1760s, was not unreasonable.[67]

Kelvin suggested that Watt should be regarded as responsible for other later engines too:

> Without a condenser, Watt's high-pressure, single-acting engine of 1761 only needed the cylinder cover with piston-rod passing steam-tight through it (as introduced by Watt himself in subsequent developments), and the valves proper for admitting steam on both sides of the piston, and for working expansively, to make it the very engine [!!] which, during the whole of the past century, had done practically all the steam work of the world, and was doing it still ... Even the double and triple and

> quadruple expansion engines, by which the highest modern economy for power and steam engines had been obtained, were splendid mechanical developments of expansion, discovered and published by Watt, and used, though to a comparatively limited extent, in his own engines.[68]

Kelvin really was drawing a long bow here! The message, however, was clear – Watt had discovered the fundamental physical principles through his experiments; all else was mechanical development. This was the sense in which Watt had been a superior engineering scientist. Kelvin ended his address by referring to the forthcoming completion of an institution, funded by the generosity of Andrew Carnegie and designed to benefit the world by scientific research – the 'magnificent' James Watt Engineering Laboratory of the University of Glasgow.

The reporter for the *Scotsman* remarked upon the way Kelvin had held the audience's attention with his 'keen and living interest ... in his subject, as for some long-standing personal friend'. Kelvin's 'eager solicitude' in presenting the life work and trials of Watt in a way that 'occasionally almost outran his physical powers' was very striking. In impressing his audience in this way, the great man indulged, from my perspective, in an orgy of presentist history of the same sort that had already, in the nineteenth century, disguised Watt the chemist as energy physicist and proto-thermodynamicist.

Conclusion

Powerful economic and social forces were involved in the varied designation of Watt as engineer. We have seen enough examples now to conclude that in the nineteenth century the character of Watt was extremely malleable, ranging from the discoverer and philosophical engineer at one end of the spectrum, through the engineer with a scientific approach, to the craftsman-engineer at the other. That character was moulded to varied purposes by different individuals and groups. The image of him as a craftsman-engineer appealed to, and affirmed the identity of, those tens of thousands of working engineers who constructed and tended the machines of the workshop of the world. Watt shared their skills, if possessing them in a highly refined degree that amounted to a true genius for contrivance. On the other hand were those who, whilst conceding Watt's genius for contrivance, found the source of his most important inventions in his philosophical character. The professionalization of academic engineering, and the need of scientific elites for practical affirmations of their importance, gave impetus to the varied construction of Watt as a 'philosophical engineer', as a 'profound engineer' and as an 'engineering scientist'. In this elevation of the profound engineer the designation of Watt as a superficial chemist was confirmed and entrenched. More than this, in a world of engineering that by the 1870s included the electrical and in which the doctrine of the equivalence of energy ruled, it was entirely appropriate that the mechanical Watt should also become – in the hands of the designators of standard units – the electrical 'Watt'. What all these characterizations have in common is that they are remarkably non-specific about what manner of science Watt's was. Such vagueness allowed the nineteenth-century hearer or reader to supply his or her own estimation, and led inevitably to the modernization of the 'Great Steamer'. The nineteenth century created Watt in its own images. By that point he had been truly cut free from eighteenth-century realities.

4 WATT'S CHEMISTRY OF HEAT

I argued in Chapter 2 that the realities of Watt's chemical work and understandings were obscured in the nineteenth century by his own narratives of his invention, and by the 'water controversy'. In the latter, elite scientists, such as members of the Cambridge group and the leadership of the Chemical Society, sided with Cavendish and condemned Watt's chemistry as archaic, more of a stain on his reputation than a valuable addition to it. Watt's supporters, on the other hand, made their claim for their hero in a way that artificially and incidentally modernized the nature of his chemical claims and arguments (as well as Cavendish's). It remains a difficult task to understand Watt's chemistry on its own terms, but in this chapter I will make an attempt. I am guided by a conviction that it is a mistake to separate in an *a priori* fashion, as authors usually do, Watt's philosophical (or theoretical) chemistry and his practical chemistry. That mistake is probably predicated also on the belief that Watt failed as a chemical theorist (that is in his public venture into the chemistry of water and airs), while still doing interesting things as an ingenious 'cookbook chemist' with glazes, varnishes, inks, bleaching agents, medically promising airs and, of course, steam.

The bifurcated approach to Watt's chemistry is misleading on two counts. First, a number of Watt's practical ventures did draw upon his theoretical ideas. Second, there is a good case for seeing Watt's chemical ideas as in turn evolving as a result of his practical engagements.[1] We will see in detail in this chapter and the next that Watt's experiments on steam, undertaken in part to inform his attack on the practical problem of steam engine improvement led him to develop a chemistry of heat that went beyond that of his mentor Joseph Black. Watt was thoroughly convinced that there was a connection between the chemistry of steam and the chemistry of airs. In fact ice, water, steam, mist, water vapour, common air and other airs were all subject to the chemistry of heat and their practical and theoretical study was interrelated as a result. This inosculation of the practical and theoretical also occurred in other of Watt's ventures. His contribution to pneumatic medicine in collaboration with Thomas Beddoes in the 1790s, for example, is usually seen as a practical one consisting mainly of improved methods for producing and delivering airs. However, as

Trevor Levere and I have shown recently,[2] Watt deployed theoretical ideas about airs in that project. Moreover, those ideas had continued to evolve and seem to have been crucially supplemented by notions about solutions garnered from the French chemist Berthollet during Watt's involvement with chemical bleaching in the late 1780s.

In what follows, then, I will try to remain open to the idea that the theoretical Watt and the 'cookbook chemist' Watt were the same person and so able to draw on each other. In all this though there is a caution to be added about Watt the chemist. Although Watt worked with, adhered to and developed the Blackian chemistry of heat throughout his active working life, he was not a systematic chemist. Nor did he take all available opportunities to mix with those who were engaged with chemistry in a more systematic and sustained fashion.[3] He appears to have read widely and operated in a magpie-like fashion, picking up bright new ideas that seemed useful to him and endeavouring to deploy them in his accounts and practices. Working in this way was probably partly an issue of temperament, but it was not discouraged by the complex, confused state that chemistry was in during the later eighteenth century. Looking backwards we can see Watt as one of many who sought to hybridize the 'old' and the 'new' chemistry. We will see, for example, that he was willing, eventually, to accept the discovery of oxygen but did not regard that as entailing abandonment of belief in 'phlogiston'. While from the position of an unwavering devotee of the new chemistry, or of the old, such a position seemed nonsensical, at the time it was a viable one, since what the new chemistry was to be was still being decided. Watt's intermediate or hybrid status as a chemist undid his chemical reputation in the long term. However, it was a position shared with others of his circle, notably Joseph Priestley, but also James Hutton, another who admitted oxygen but retained phlogiston. Even Erasmus Darwin, who is usually seen as a convert to the Lavoisian scheme, is revealed when examined closely as only partially so. He too was a hybrid, though one closer to the Lavoisian end of the spectrum than Watt, Priestley or Hutton.

In treating Watt and others as hybrids in this fashion, however, there is a danger of missing what is probably a deeper divide between the defenders and the opponents of phlogiston. We might consider those concerned with the chemistry of airs in the later eighteenth century as divided among 'lumpers' and 'splitters'. The splitters were inclined to see the new airs being discovered as having distinct chemical identities. The lumpers, on the other hand, whilst they might admit that there were distinct airs (and even that the population of known airs was increasing) still believed in the fundamental unity of various families of airs. Watt and Priestley were lumpers, Lavoisier and exponents of the new chemistry were splitters. Richard Hills presents Black as a splitter but, as we will see, he could also be readily characterized as a lumper.[4] Lumpers appealed to unifying agents. This might be phlogiston or heat, the variability of airs being due to the amounts of heat combined with a 'base' of some sort. The unifying agent might be dealt with as a 'principle'. De Luc's use of the idea of 'humor' as the principle of humidity – which we will come across because Watt adopted it – was another example of a lumping device. The example of 'humor' – the common basis of air, water, and water vapour – also shows how lumping was a tendency across other boundaries that we tend to assume the existence of. Thus the boundary between air (gas) and vapour (for example steam) was recognized in the sense that the former would not condense, but they were also lumped together as 'elastic fluids'. Their differences might be overcome if extreme enough conditions of temperature and pressure were available. Airs would condense if low enough temperatures could be reached. Water, and vapours, would become airs if high enough temperatures (and low enough pressures) were involved. In this sense a chemistry of principles still had a strong pull on the minds of many pneumatic chemists.

I will begin by investigating Watt's chemical apprenticeship with Joseph Black and the ideas and practices that the young Watt would have been exposed to. This is the context of his first steam experiments. A second phase of Watt's chemical work came after his removal to Birmingham and involved a very close working relationship with Joseph Priestley. This was the period of the publication of his ideas about the chemistry of water. Finally I will look at Watt's evolving chemical ideas from the late 1780s into the early 1800s.

Watt's 'Apprenticeship' with Joseph Black

The problem with much of the work on Watt's inquiries into heat is that it fails to take seriously the nature and the lasting consequences of Watt's early natural philosophical beliefs. We need to be alert to what Watt *could* think between 1760 and 1800 when he was active in relation to the steam engine, and to be so alert we must understand his relationship with the great chemist Joseph Black, and with the tradition to which Black belonged.

Black and his circle present a historiographical problem that has never been tackled so far as Watt is concerned. However, some time ago historians began to confront similar problems for certain of Watt's friends and contemporaries. James Hutton is an interesting case, not least because his famous theory of the earth was wrested from context in a way similar to the fate of Watt's work on steam. Restoring the appropriate context in Hutton's case also involved understanding the importance of his chemical theory of heat. Patsy Gerstner argued back in 1968 for the importance of Hutton's theory of heat to his theory of the earth.[5] Arthur Donovan subsequently explained why this fact was given little prominence in the accounts of Hutton's theory developed by his disciples, notably John Playfair and Sir James Hall. As Donovan characterizes the problem: 'to perceive the origins of Hutton's theory, and more particularly its relation to the older chemical tradition, we must first penetrate the historical smoke screens thrown up by the revolution in chemistry and the birth of geology'.[6]

Between the 1750s and the 1790s a group of Scottish natural philosophers, of which both Hutton and Watt were a part, pursued a 'chemical theory of heat' – that is, in Donovan's terms they treated heat as a material substance involved in chemical processes. Their ideas were rooted in the Boerhaavian tradition and the ideas of other continental chemical natural philosophers, partially mediated by William Cullen, treating heat, or fire more generally, as chemically responsible for the state of material bodies. Thus Hutton's theory of forces used the ideas of specific and latent heats but in a way that looked back to the earlier days rather than forward to their subsequent meanings. Hutton deployed specific and latent heats as parts of the 'repulsive force' or what he called the 'solar substance'. For Hutton, specific heat was a 'principle'[7] influencing volume, latent heat was the principle of fluidity. Together with light and electricity these made up the repulsive force. So for Hutton latent heat causes fluidity and specific heat causes expansion. The dynamics of natural cycles of rock formations are largely explained in these terms within his theory. It is for this reason that Donovan claims that Hutton's theory of the earth is given its dynamics by Hutton's chemical theory of heat.

Hutton's theory was subsequently reinterpreted and reshaped to find honour and application in a world where a recent revolution in chemistry and the ambitions of the exponents of an independent science of geology were powerful intellectual forces. That was how Hutton's chemical theory of heat disappeared from his theory of the earth as published and expounded by his followers. For them the chemical theory of heat was an excisable irrelevance and embarrassment.[8] Similarly, I argue, the intellectual foundations of Watt's chemistry (set in place in the late 1750s and 1760s, and developing with the work of his friend Black) and their importance to his steam-engine work, have been obscured by both the chemical revolution of the late eighteenth century and by seeing Watt's engine work through the lens of subsequent theories of heat engines and thermodynamics. So let us turn now to Watt's 'apprenticeship' to Black.

That Watt did serve an informal chemical apprenticeship under Black is certainly true, if distressingly hard to document in real time. There is little documentary evidence of their relationship before Black moved to Edinburgh in 1766. However, there was much more to their relationship than master–apprentice. In their edition of the letters of Watt and Black, Robinson and McKie painted an influential picture of the relationship between the two men as 'Partners in Science'. Yet their perspective on the relationship drew not only from the marvellous variety of the (post-1766) correspondence but also from the vexed historiography of the partnership that the nineteenth century had generated. Thus, overall, Robinson and McKie characterized the relationship as an 'association between the most eminent chemist of Europe, which meant, at that time, the world, and Europe's greatest engineer'.[9] Approaching their relationship in this way is to prejudge too much. We become predisposed to regard Black as inhabiting the world of 'theory' and Watt that of 'practice'. We are encouraged to see Black as a cosmopolitan figure of the Scottish Enlightenment and Watt as a clever engineer hanging from his coattails, their relationship involving a 'transfer of intellect to craft'.[10] This is a mistake.

There is a case to be made that in some respects Watt was, through much of their association, more of a chemist than Black. The professor made his key discoveries early in life. His move to Edinburgh in 1766 was, as Donovan tells us, 'the end of his career as a creative philosophical chemist'.[11] Black was remarkably reticent about publication – Watt and others constantly cajoled him to claim what was his – and he exerted his primary influence orally, as a teacher of generations of students in Glasgow and then in Edinburgh.[12] A proportion of those students made significant contributions to the science of chemistry themselves and often acknowledged a great debt to their teacher, but that teacher was himself remarkably absent from the public 'conversation' and the great debates occurring in chemistry in the late eighteenth century. Watt did not publish a great deal either and, as we have noted, his chemical work was episodic and undisciplined, but he collaborated not only with Black but also with Priestley, DeLuc and Beddoes, and he had some claim to the discovery of the compound nature of water. He met, and discussed the chemistry of the day with, Lavoisier, Berthollet, Laplace and many others, wrote a circulated response to the new chemical nomenclature proposed by Hassenfratz and Adet, and was known as a defender of phlogiston theory. In short, Watt might reasonably be regarded as a more active chemist than Black in the later decades of the eighteenth century. We will see that in important respects he did more than Black ever did to explore the wider chemical ramifications of the chemistry of latent heat.[13] Having said this, we must remember that for Black the 'chemical life' was focused through the relationship that he had with his students. They became the key vehicle for the communication of his ideas. Black had taken up the example set by William Cullen in this respect. Just as Cullen had cultivated his relationship with students, some of whom became almost collaborators (including Black), when they were invited into the laboratory, so Black did too. Robison and William Irvine (as students) and Watt (on a different basis) became experimental apprentices to Black within this tradition.[14]

The distance between Black and Watt is also reduced when we recognize that Black had an entrepreneurial side, or at least an interest in funding practical, potentially profitable projects. The first partnership between Black and Watt was in late 1759 when Watt, Alexander Wilson and Black associated in a venture to produce improved clocks. Other partnerships involving the two men (sometimes with other parties) included a hardware shop and a scheme for producing alkali from salt.[15]

When Watt moved his business into the precincts of Glasgow University at the end of July 1757, he had a number of connections with the professorate there, especially with John Anderson, Professor of Natural Philosophy from 1757, and Alexander Wilson, who became Professor of Practical Astronomy in 1760. Black was appointed to the Chair of Medicine at Glasgow in early 1757. Few records of this early period of their relationship survive – understandably, the all-important letters between them only begin in earnest after Black's move to Edinburgh in 1766. We do know that at Glasgow Watt was repairing lecture equipment for Black in October 1757 and undertook other commissions for him.[16] Robison, whose own association with Black and Watt began the following year, later suggested that Watt 'was much with the Dr. during his most important researches' (that is, Black's researches on heat which were conducted in the early 1760s), but the truth of this is unknown. If we are to believe Watt's claim that he only learned of latent heat from Black after he had conducted his own experiments revealing the phenomenon, then it seems unlikely because Watt's own experiments post-dated Black's. At the very least, though, and as Watt repeatedly acknowledged, he did acquire much knowledge from Black (though through informal contact rather than formal study) and learned from him 'the correct modes of reasoning, and of making experiments'.[17]

Chemistry was to change a great deal during Watt's lifetime, but his basic approach to the subject was laid down in the late 1750s and early 1760s during his association with Black. In his turn, Black had imbibed the approach and interests of his predecessor William Cullen, with whom he had studied and worked as an assistant. So far as the study of heat was concerned, Cullen's key publication appeared in 1756.[18] For him, as for Black and Watt at this time the basic form of chemical inquiry was still set by Hermann Boerhaave's *New Method of Chemistry*, in which the field was described as the manipulation of the four elements in menstruums contained in vessels.[19] Black's instrumental armamentarium included furnaces (which Watt helped to repair, redesign and rebuild) and an array of ordinary household equipment, including kettles.[20] Where Black differed in approach from Boerhaave, according to Cardwell, was in not concerning himself much with the metaphysics of causal agents, notably 'fire'. Instead, Black developed quantitative investigations of operationalized concepts of temperature and heat.[21] This set the basic format of Watt's chemistry too. From the beginning, his experiments on steam were quantitative in this same fashion. While Cardwell's identification of this form of chemical investigation is certainly correct, it is a mistake, I think, to argue from it that Black was a 'positivist'. In making this claim, Cardwell and Hills (and also Donovan) suggest that Black approached the study of heat in a way that presaged nineteenth-century thermodynamics. Hills, drawing on Donovan, put it this way:

> in dealing with heat, Black was looking for general laws which would explain changes in the states of matter and was not trying to suggest a new theory of heat. He applied general laws of heat, flow, change of state and so on to particular cases dealing with specific substances and in this he had much more affinity with nineteenth century physicists who developed the study of thermodynamics than with his contemporary fellow chemists.[22]

It is true that Black did little himself to develop the consequences of the chemistry of heat for the chemistry of gases and of other substances. But that does not mean that for him heat was a physical externality to chemical process. On the contrary, as Golinski puts it:

> Although Black imported thermometric techniques into chemistry from other physical sciences, to describe his view of heat as "physical" would be inaccurate. Rather as chemistry was understood in terms of heat, so heat was understood by Black in terms of chemistry. As Robison clarified, Black perceived the latent presence of heat in bodies in terms of its chemical combination with ponderable matter...[23]

The fact is that Black failed to put his conception of latent heat to work in accounting for other chemical phenomena, whether because his energies were diverted from novel investigations or because of his much vaunted reluctance to build systems. In this sense his lectures on heat remained peculiarly disengaged from the rest of the substance of his chemical lectures. Nevertheless it was the chemical combination of heat with other matter that explained for Black the nature of fluidity. Even as he shared the Boerhaavian conception of chemistry as the operations of heat, Black moved away from Boerhaave's treatment of heat as an external 'instrument', embracing the idea that heat participated in chemical combinations.[24] Having sided with Golinski (and against Cardwell and Donovan) on this point, I would nevertheless concede that defending either position is very difficult because of the lack of relatively unmediated access to Black's work. Because of his reluctance to publish, we see his work only through filters provided by others, especially Robison. Those filters were activated in part by events subsequent to Black's experiments and discoveries, by solicitude for the reputation of a departed friend, by the local and international politics of science, indeed by international politics as such.[25]

These contrasting historical interpretations of Black mirror the different ways in which his 'disciples' developed the chemistry of heat, and of latent heat in particular. Thus William Irvine's non-chemical theory of latent heat did indeed take Black's ideas into a form readily assimilable to the early nineteenth-century transformation of heat into a *physical* instrument of chemical change. In drawing lines from Black to proto-thermodynamic thinking, 'Irvinism' was the conduit. This is true, I think, even though, as Robert Fox long ago showed, Irvinist conceptions of heat were under apparently fatal attack from French investigators after the first decade of the nineteenth century. Although Fox, with some reason, finds the rearguard defence of Irvinism to involve 'muddled thinking' in a situation of 'hopelessness' by the 1820s,[26] its doctrines had already generated a view of Black's achievement that swept to one side the importance of his chemical theory of heat, adding to the real historical force of Robison's attribution of a positivist approach to Black in his publication of Black's *Lectures*. It was these developments, rather than those in France, which were just beginning to repudiate Irvinism directly, that were uppermost in Watt's mind in the very early 1800s as he refashioned his own experiments on steam for posterity. At this juncture a little more needs to be said about Irvine.

William Irvine was a native of Glasgow, matriculating at the university at the age of thirteen in 1756. Black soon noticed his interest in chemistry and his disposition to apply mathematics in his studies. Like Black, Irvine's main lines of communication for his work were through his students and the occasional excursion before the local Philosophical Society. Working as Black's assistant, Irvine helped in his professor's determination of the latent heat of steam and contributed values of that quantity for melting tin, zinc and spermaceti and beeswax. He also worked with Black on establishing experimentally the specific heats of various substances. However, Irvine became very much his own man. Andrew Kent puts it colourfully:

> Irvine ... laboured long and effectively in support of his teacher's enquiries; but following Black's departure for Edinburgh, this spagyric Man Friday showed a stubborn originality in the interpretation of their results. He found Black's theory of Latent Heat unwelcome for various reasons. It presumed a material view of Heat where the younger man was anxious to preserve an open mind: it ascribed a duality of character to Heat as 'free' or 'combined', which Irvine was anxious to avoid. Black agreed that the doctrine taught by his successor was "an ingenious thought", but genially maintained his own opinion.[27]

Irvine's 'ingenious thought' was that latent heat was the difference in heat content of two phases of a substance at the same temperature. Thus, at the moment when a substance moved from solid to liquid its heat capacity increased instantaneously. This, according to him, rather than any process of chemical combination of heat with the substance, explained the extra heat required to support a change of phase. Irvine also used considerations of specific heat to derive a value for absolute zero of temperature. The fact that Watt defended Irvine's priority in this against other claimants assures us that Watt was entirely familiar with Irvine's ideas.[28] Watt defended Irvine, as well as Black, against those who, like Adair Crawford,[29] appropriated his ideas. In about 1769 Irvine and Watt worked together on 'manganese' as a colourant in glass manufacture. They also worked together on heat – many of the tabulations of results from Watt's experiments in the early 1770s on the latent heat of steam at different pressures are recorded in Irvine's hand.[30] However, for all his respect for Irvine, Watt's development of the Black tradition, like James Hutton's, unambiguously embraced the chemical conception of latent heat that Irvine rejected.

For Watt as for Joseph Black, the issue of the chemical nature of water, steam and air was closely tied up with the understanding of heat. Their laboratory work at Glasgow in the early 1760s was concerned in various ways with the heating of water, and especially with what we would call 'changes of state', a notion that we must be very careful not to attribute to them since they did not share our conception of a 'gaseous state'. This becomes clear when we realize that for them the composition and decomposition of water was to be conceived of as involving the same sorts of processes in the changing of water into air and air into water. Watt's work on the composition of water, on the one hand, and his experiments on steam in relation to his engine improvements on the other, have been treated separately by historians, but I suggest that Watt himself, and his friends and associates made no such clear distinction – there was an essential continuity between them.[31]

In his early Edinburgh lectures Joseph Black defined chemistry as 'the Study of the effects of Heat & Mixture on Bodies and Mixtures of Bodies, with a view to the improvement of our knowledge'.[32] Black's lecture courses in chemistry always began with the topic of heat and, in fact, struggled to leave it behind. Black's first chemistry lectures were delivered in Glasgow in 1757–8, and his association with James Watt, who had set up shop as an instrument maker within the University of Glasgow, dates from this time also. Black's work on heat was partly inspired by the experiments of his predecessor as Professor at Glasgow, William Cullen, on evaporative cooling and partly by meditations on processes of freezing.[33] Black became convinced that processes of freezing and vaporization involved large quantities of heat well beyond those that ought to be required according to prevailing ideas. His first major experiments on raising the temperature of water and ice were conducted in 1761, and a year later he performed experiments on vaporization. From these experiments, values were obtained for the quantity of heat required for (what would later be called) latent heat of fusion and of vaporization.

In the winter, spring and summer of 1764 Watt performed a series of experiments of his own on steam. He first became involved in this work through his famous efforts to repair a model of a Newcomen engine undertaken as part of his duties as instrument maker to the University of Glasgow. He was able to show that steam could heat about six times its own weight of water to boiling point. In working that model engine Watt was struck by the surprisingly large quantities of cold water needed to condense the steam in the cylinder. Watt mentioned this result to Black, who then explained the idea of latent heat. The precise relationship between the concept of latent heat and Watt's steam engine work is very controversial, as we have seen. My immediate concern, however, is to emphasize the fact that from 1764 Watt's work on heat was undertaken as part of Black's tradition of work. Black always made prominent reference to Watt's experiments in his lectures at Edinburgh University so far as we can rely on Robison's edition of Black's lectures.[34]

Black's departure for Edinburgh in 1766 obviously reduced his personal contact with Watt, but they corresponded with reasonable frequency. The chemical aspect of their correspondence is often of a very practical kind, with Black asking for equipment or requesting specimens. But the topic of heat lay close to the surface. Practical matters could easily raise important theoretical issues. For example, when Watt ventured into the production of Delft pottery he conducted lengthy experiments with different designs of kiln, with the use of coke instead of coals to fuel the kiln, and also with glazes of various sorts. He sought Black's help and reassurance on many of these matters. Informed of Watt's puzzling over the uneven heating of a kiln fuelled by coke, Black reminded him that whilst the flames of coals represented the flow of heat, coke communicated heat in a way more like the heat from the sun. Black reminded Watt of Nollet's experiment with mirrors, a glowing piece of charcoal and a sample of gunpowder, in which the latter was ignited by communicating the heat of the glowing charcoal from the focus of one mirror to that of the other. Watt responded to being alerted to what he called the 'beaming' of the glowing coke by redesigning the kiln to incorporate a metal grid that would in effect relay the beaming of the heat and ensure a more even heating of the kiln.[35] This is a good example of their theoretically informed and guided practical work. The systematic unravelling of much of Watt's cookbook chemistry would, I think, reveal similar strategies and processes often at play. Although Watt never systematically laid out his ideas on the composition and firing of different clays and glazes, for example, it would be surprising if there too his ideas about heat as transformative agent were not involved. We have yet to discern, I suspect, many of the underlying theoretical threads of Watt's cookbook chemistry.

After Watt moved to Birmingham in 1774 he remained loyal to the Blackian tradition. Thus, for example, in 1779 he reported to Black that he was standing firm against those who challenged him with a non-material theory of heat:

> I am most sadly teased with the philosophers in this country and their theorys of heat being only a vibration in bodies which seems to me highly improbable and not to be reconciled to many notorious facts – I see Macquer has defended it ably but has brought no conviction to me – I think that latent heat and difference of equilibrium will account for all its phenomena except the heat produced by friction there I am at a loss.[36]

A few years later, when Lavoisier and Laplace's memoir on heat was published, Watt ensured that Black received a copy and told his friend that the piece was well-written 'though not free from objections'. He lamented the fact that the authors did not mention Black's name, though Black was himself partly to blame for this by not publishing. Nevertheless Watt believed that Black's ideas would prevail: 'I think that as far as I can see into the matter that Dr. Irvine's doctrines and Dr. Crawfords of capacity, will fall to the ground, and your original theory of Latent or essential heat be established'.[37]

For Watt and Black, then, the phenomena of heat were integral to *chemical* processes. David R. Dyck in an unpublished PhD thesis gave a very useful account of Watt's chemical work in which he made a number of very important points about the connection between Watt's work on steam and that on water.[38] Dyck discerned that Watt's explanation of the composition of water had in part grown out of ideas about the relationships of water, air and fire that existed in older traditions. Watt long believed that if water could be heated enough it would change not just into steam but ultimately into air. We will see that this idea had proximate causes for Watt, but in its general form, Dyck ventures, Watt could have acquired it from the German medical man and chemist Johann Theodor Eller[39] who suggested in 1746 that water is converted into air by combination with elementary fire. As Hélène Metzger long ago discussed, the early decades of Watt's interest in chemistry and heat (the 1750s through the 1770s) saw a proliferation of ideas about entities actively involved in chemical reactions including not only phlogiston but also heat (fire) and light.[40] Herman Boerhaave, the great Dutch physician and chemist had identified fire as the source of fluidity in water. Eller was a student of Boerhaave as was Cullen. Boerhaave's influence in Scottish medicine and chemistry was very strong. So whilst we lack direct evidence that Watt imbibed his ideas from Boerhaave and Eller via Cullen and Black, there is considerable circumstantial evidence to suggest that this was so. Watt read widely and was well aware of Boerhaave's *New Method of Chemistry* and other influential chemical works.[41]

Watt, like Black, was committed to one of three major views of heat extant at the time. The first of the three views was that heat was motion, or the vibration of the parts of ordinary material bodies. This mechanical theory of heat had been favoured by Boyle and had been endorsed by Newton. But the mechanical theory was not fashionable in the mid- to late eighteenth century. We know that a mathematical theory of heat as motion was developed by Henry Cavendish in the 1780s but, typically, not published.[42] This type of theory was, of course, to become the 'correct' view of heat by the mid-nineteenth century. The second and third accounts of heat are often collapsed together as 'material' theories since in both heat was a special substance rather than the motion of ordinary matter. The distinction between these two material theories is clearly described by McCormmach:

> In one version, when fire is in a free state, its particles move rapidly in all directions, their motion registering as sensible heat; the particles lose their motion when they combine with particles of ordinary bodies, in which state fire loses its property of sensible heat. In the other version, the particles of fire do not enter into combination with those of ordinary bodies, and fire always retains the property of sensible heat; because different bodies have different capacities for heat, an addition of the same quantity of fire results in different increases in sensible heat.[43]

Joseph Black subscribed to a version of the first material theory. He considered heat to be a special form of matter that combined with ordinary matter as a result of chemical forces of attraction. The phenomenon of cold produced by evaporation was explained by Black as follows. The cold experienced when water evaporates is the result of the water absorbing sensible heat as the water becomes vapour. The heat is not lost, rather the heat combines chemically with the vapour and it is this that gives the vapour the property of elasticity. Thus water absorbs the matter of heat, which becomes latent (or 'fixed' as Black termed it initially) because it is now chemically combined with the water. Through this combination, the latent heat confers the property of fluidity or elasticity upon the vapour, that is upon the steam.

This account seems strange to us. We are accustomed to thinking of Black as 'the discoverer of latent heat', and by this we usually mean the discoverer of what we understand by the term 'latent heat'. But this is quite wrong. Our concept of latent heat is underwritten by a kinetic theory of heat; Black's was based on a material (chemical) one in which latent heat was intrinsic to chemical reactions and was compounded with ordinary matter. The fact that Black has a conception of latent heat 'foreign' to us is made clear by the terms in which Black's supporters stated and defended his claims. Thus, in his *Encyclopaedia Britannica* article on 'Steam' Robison described Black as having 'justly concluded' that,

> water differed from ice at the same temperature by containing as a constituent ingredient, a great quantity of fire, or of the cause of heat, united with it in such a way as not to quit it for another colder body, and therefore so as to not go into the liquor of the thermometer and expand it. Considered therefore as the possible cause of heat ... if any more heat was added to the water it was not latent, but would readily quit it for the thermometer, and, by expanding the thermometer, would show what is the degree of this *redundant* heat, while fluidity alone is the indication of the *combined* and latent heat.

Once again, Robison, in defending Black against what he took to be De Luc's unjustified pretensions to have made the discovery, stated that Black was 'the first Man *who had conceived Heat or Fire as a constituent part of all Fluid and Gases*'.[44]

We can now begin to appreciate what Watt meant when he discussed water, steam and air. Watt characterized his long-held views on the relations of water and air in a letter to Matthew Boulton in 1782:

> You may remember that I have often said that if water could be heated red-hot or something more, it would probably be converted into some kind of air, because steam would in that case have lost all its latent heat, and that it would have been turned solely into sensible heat, and probably a total change of the nature of the fluid would ensue.[45]

Watt expressed the same idea in a letter to Black at the same time.[46] Writing about what he saw as his own contributions to their joint efforts, Watt explained:

> What ... I think my own, is the trying the experiment on the latent heat *in vacuo* and the finding it to be greater than under the pressure of the Atmosphere – The exp[erimen]ts to ascertain the different degrees of heat at which water boils under different pressures – the expansion which steam in its perfect state receives from heat and the experiments on the bulk of water when converted into steam together with a theory which I have devised which accounts for the boiling heats of the water not following a Geometrical progression, and showing that *as steam parts with its latent heat as it acquires sensible heat or is more compressed, that when it arrives at a certain point it will have no latent heat and may under proper compression be an elastic fluid nearly as specifically heavy as water and at which point I conceive it will again change its state and become something else than steam or water. My Opinion has been that it would then become air, which many things had led me to conclude...*[47]

When Watt published on the composition of water in the *Philosophical Transactions* he was to drop this specific idea. However, even as he suggested there that water is produced by the sparking of inflammable and dephlogisticated airs, in a way that involves latent heat less directly, he still retained a central *chemical* role for heat in the process.[48]

By the time of this accounting of the joint efforts of Black and himself, in 1782, Watt's chemical interests had been rekindled after a period of desuetude in the later 1770s when he was preoccupied with the hands-on management of the steam engine business, especially in Cornwall. But by 1780 the most difficult years were behind Boulton and Watt, and Watt was beginning another phase of chemical work during which he entered new collaborative relationships. However, his communications with Black continued and were occasionally revealing of their chemical relations.

After Black's death in 1799, his family raised the issue of the publication of his famous lectures. At first there was some confusion about who would undertake this task. Adam Ferguson, Black's obituarist for the *Transactions* of the Royal Society of Edinburgh, was willing to produce a straightforward transcript of Black's lectures. John Robison was also approached and declared himself available, though he believed that Watt was the ideal man to do it.[49] He put it to

George Black Jr that 'if Mr Watt would undertake it, there is not a Man living who can do it so well, or do it with more delight. He is my superior in chemical knowledge and Experience, and was continually with your Uncle during his most original Researches'.[50] Watt refused to take on the task but agreed to meet Robison in Glasgow to consult on the transcription of the manuscript lectures. Robison still wanted to make use of Watt as the best person to help with interpretation of obscurities in the manuscripts. The plan to publish a 'straightforward' transcription could not survive an examination of the manuscripts, which were disorganized and written for oral delivery. Robison was also concerned about their lack of art or literary merit. Black had aimed to be understood by the least, and the least cultured, of his students. Transcribing such lectures would not do justice to Black's memory, Robison thought. They contrasted starkly with the artful, systematic texts of the French chemists. The lectures were also dated. Black had interpolated comments over the years on developments in the field, but these were tacked onto an unchanging structure. Robison came to the view that Black's interest in the cutting edge of chemistry had faded over the years. He had failed, because of this and because of declining health, to respond in any substantial, or critical fashion to the chemistry of Lavoisier. Robison was disappointed to see that Black had uncritically imported swathes of French chemistry into his lectures, which sat incongruously beside more traditional ideas.[51] This too led him to despair of the published lectures living up to Black's great name. He was inclined to publish instead a compilation from Black's important discoveries in a way that would do greater justice to his memory. In the end Watt made a suggestion that seemed a reasonable and viable compromise. As Robison explained Watt's idea: 'He considers it as a *History of Chemistry*, for forty years, by one of its greatest Masters'. The lack of system was true to Black's suspicion of intellectual systems in general – the deficiency could be made a virtue of and even elevated to the status of a methodological precept.[52]

Black's *Lectures*, as edited by Robison, have been the source of much of the historiographical confusion about the relationship between Watt and Black. As we have already seen, Watt did not like the way that Robison ended up portraying his dependence on Black. Watt's displeasure came to Robison's notice as soon as Watt became aware that the *Lectures* had been dedicated to him as a *pupil* of Black. Watt apparently repudiated the idea to others in conversation, which Robison learned about:

> A Gentleman told me that he [Watt] said in a company of several Persons that he never was a Pupil of Black, I am therefore mistaken as to that fact – of which I never doubted, having found him, on my return to College in 1763, most intimately connected with Dr Black, and continually speaking of his lectures and doctrines – possibly I took it for granted without its having been told me – But, even granting

> the mistake, as to his having regularly attended the Lectures, I must still consider Mr. Watt as Dr. Black's most illustrious Pupil...[53]

What we can say for certain is that Watt owed a great deal to Black for his initiation into the science of chemistry. Watt took further than his master the commitment to a chemical theory of heat and the working out of its consequences. Watt took more seriously than his master the task of deciding what was useful in the new French chemistry and what had to be resisted. His conception of heat-content as a key differentiator between different chemical substances, such as different kinds of airs, grew out of these early experiences. So too, did the link in his mind between the chemistry of steam and the chemistry of airs, the idea of the essential unity of the chemistry of elastic fluids.

Watt, Priestley and the *Philosophical Transactions* Papers

From about 1780 James Watt's interest in chemistry revived strongly, and did so, I would claim, in connection with further steam engine innovation. In retrospect Watt has often been seen as a conservative innovator. There are a number of reasons for this perception, but a major one is his resistance to the idea of working engines at higher pressures. However, there are many signs that having got his first engines running Watt was in fact very open to the possibility that steam-engine design might change radically. Some potential changes, like the possibility of using alternative airs as a basis for novel engines, he took seriously as a potential threat, at least for a while. Watt's correspondence with Matthew Boulton at this time reveals that Joseph Priestley conducted experiments on their behalf to investigate the possibility of deriving mechanical power from the condensation of air, or of using in an engine any of the new airs recently discovered. The concern that others, notably the Hornblowers, might get a march on Boulton & Watt in this respect, just as their engines were beginning to pay, was soon dispelled, however. Boulton explained to Watt that he was 'persuaded from Expm^{ts} made by Dr. P. that non of his Airs will ever incommode us ... nothing is to fear from any of the tribe of Gasses'.[54] Other possibilities for innovation lay in managing heat and steam differently, and Watt was more convinced than ever that knowing as much as possible about steam, and about airs, was crucially important. Watt's intuition, I think, was that the understanding of steam was to be gained at the intersection of Black's chemistry of heat and the chemistry of airs that Joseph Priestley had brought into his ambit in a newly-concentrated way when the radical divine arrived in Birmingham in 1780. Although, as we have seen, Watt remained in sporadic contact with Black, and continued to defend Blackian chemical views, Priestley was to be the touchstone of Watt's chemical life from that time onwards.

Priestley had settled in Fair Hills Birmingham sometime in September or October 1780.[55] In establishing himself there, he had financial support from Boulton and Darwin to help build and equip a new laboratory. Priestley was, of course, already very well known for his scientific discoveries. A number of the active members of the Lunar Society had corresponded with him, or discussed his work. Watt, for example, had shown great interest in 1773 in Priestley's discovery of nitrous oxide and his ideas about photosynthesis.[56] Priestley was readily accepted as a member of the Lunar Society. In fact the Society's day of meeting was changed from Sunday to Monday to accommodate his duties as a minister.[57] In recalling those days, Priestley considered his settlement at Birmingham as 'the happiest event of my life'. His philosophical objectives were well served, he recalled, by the presence of 'good workmen of every kind, and the society of persons eminent for their knowledge of chemistry, particularly Mr. Watt, Mr. Keir and Dr. Withering'.[58] The fact that Priestley coupled Keir and Watt in this way is perhaps telling. Priestley had earlier asked Keir to read over his work, explaining privately to him that he was 'afraid of tripping on chemical ground'.[59] Priestley regarded Keir as having chemical expertise beyond his own. Whilst it might not match Keir's, Watt's status as a chemist in the estimation of his new collaborator was also clear.

The new 'lunatic' showed great willingness to assist in the practical ventures of his fellows. Even if Priestley had not been inclined to help Boulton and Watt, Boulton's patronage gave some leverage, which he was prepared to use if necessary. Thus, in relation to experiments on the possibilities of deriving mechanical power from the repeated condensation and vaporization of airs derived from fixed alkali, Boulton wrote to Watt:

> I have placed a confidence in Dr Priestley who hath seriously promised me to keep the secret & to give me all the facts and all he can discover in that line. The D^r is under some obligations to me that you know not of, & I think will not betray me, I am sure he will not.[60]

For his part, Watt was more unconditionally welcoming and Priestley became part of the Watt family circle. They were to experiment together on numerous occasions. For example, Priestley collaborated with Watt and De Luc in experiments on heat, steam and evaporation in 1783, and there are indications that before and after this Priestley joined Watt at home in experimental work.[61] They also routinely shared news of each others' ideas and experiments. Very early on, Priestley acknowledged the value of Watt's comments on his work in volume 2 of his *Experiments and Observations on Natural Philosophy,* and a brief letter from Watt, among others, commenting on aspects of the content of that work was included in the Appendix added after Priestley's arrival in Birmingham. Given that the preface of *Experiments and Observations* was dated 24 March 1781, it is clear that Watt and Priestley 'hit the ground running' in their collaboration.[62] Some of their ideas were perhaps converging at that time. Priestley noted in *Experiments and Observations* that in his first volume he had 'supposed heat to consist in a vibratory motion of the particles of bodies'. He still believed this in 1781 so far as sensible heat was concerned but conceded that 'there may be a *principle of heat* latent in bodies, and not manifest by any sensible effect. Heat may therefore be what is usually termed *a substance*'.[63]

One of the main outcomes of Watt's chemical activities in the early 1780s was his only high-profile chemical publication, the paper offering 'Thoughts on the Constituent Parts of Water and of Dephlogisticated Air', which was published in the *Philosophical Transactions* of the Royal Society in 1784.[64] This paper is, as we have seen, the chief basis for Watt's claim to have discovered the composition of water. Whilst its importance in that connection is undeniable, discussion of it in relation to the water controversy has, I believe, contributed to the neglect of serious attempts to understand the complexities of the chemical views that it presents.

This narrowness of focus has been engendered by a number of preoccupations of historians as they have examined the paper. One such concern is the question of priority of discovery – whether Watt, Cavendish, Priestley, Lavoisier or some other candidate should be awarded it. As a result Watt's famous words in which he summarized his view of the composition of water have been extracted from the paper numerous times. Those words have been treated as isolated speculation – a bright idea suddenly arrived at – inspired by Priestley's experiments on sparking dephlogisticated and inflammable air. The overall argument of the paper, and its intricacies, have been neglected. Watt had much more to say – in fact most of the paper (eighteen of twenty-five pages) was concerned with a sequence of experiments on dephlogisticated air, many of which Watt performed himself. Moreover, these experiments are not just additional matter. Once properly understood, they change the meaning of what Watt had to say about water. His was no simple, prescient flash of insight that water is compounded of two airs. It was an exploration of the relationship between air, water, earths, acids and heat. It did not treat dephlogisticated and inflammable air as elementary building blocks from which water was constructed, rather it invoked (as can be seen when we look closely) a chemistry of principles, centred on Jean André De Luc's notion of 'humor'.[65] Although Priestley was an inspiration and sounding board, the work of many others is considered and reflected in Watt's writing, including, besides De Luc, Richard Kirwan, and Lavoisier and Laplace. The paper represents, then, a curious unstable synthesis of Watt's and Black's work on latent heat of steam, Priestley's and Watt's investigations of airs, and De Luc's concept of humor which came from his meteorological writings on evaporation.

The contentious story of the paper's production has also distracted from serious consideration of its substance. It began life as a letter from Watt to Priestley written in April 1783 and intended to accompany Priestley's report to the Royal Society of his original experiments on the seemingly direct production of air from water in an earthenware retort. That letter to Priestley circulated in London amongst Royal Society circles before Priestley retracted his claims about the experiment and Watt withdrew his letter. From these events stemmed the subsequent claims (from the Watt camp) that Cavendish and Lavoisier, both with the aid of Charles Blagden, plagiarized Watt's ideas. Watt reworked his paper, removing the focus on the earthenware retort experiments, and presenting it this time in the form of a letter to De Luc. This letter, as published, is rendered a strange production by its consistent, extensive and explicit quotation from the earlier letter to Priestley – a device adopted because of the issues of priority that haunted its production. In what follows, I try to avoid the narrow focus that has hampered prior treatments of this paper. I consider it in its entirety, trying to appreciate its intellectual context and, by so doing, hope to advance our understanding of Watt's chemical thinking.

Watt's first observation concerned the nature of inflammable air, which Priestley, and before him Kirwan, had noted as seemingly identical to phlogiston – Priestley suggested that it was pure phlogiston, Kirwan that it was phlogiston in its gaseous state. For his part, Watt essentially concurred but believed that inflammable air also contained 'a small quantity of water and much elementary heat'.[66] The experimental basis for this belief lay in Priestley's experiments on heating metallic calces in an atmosphere of inflammable air only. The fact that the calces were reduced to their metallic state and that the inflammable air was consumed in the process led to the conclusion that inflammable air must be, primarily, phlogiston. Watt then proceeded to consider the famous experiments in which dephlogisticated and inflammable air were sparked together:

> The first effect was the appearance of red heat or inflammation in the airs, which was soon followed by the glass vessel becoming hot ... and as the glass grew cool, a mist or visible vapour appeared in it, which was condensed on the glass in the form of moisture or dew.[67]

The moisture was water and found to be 'equal in weight to the airs employed'. That the airs had been consumed in the process was shown by the fact that, the containing vessel having its mouth placed under water or mercury, the vessel was almost completely filled with the fluid. So what had happened here, according to Watt, was that the two kinds of air 'unite with violence, they become red-hot, and upon cooling totally disappear'. The only products are water, heat and light, the only product remaining at the end of proceedings being the water. Watt then offered the famous, oft-quoted words:

Are we not then authorised to conclude, that water is composed of dephlogisticated air and phlogiston, deprived of part of their latent or elementary heat; that dephlogisticated or pure air is composed of water deprived of its phlogiston, and united to elementary heat and light; and that the latter are contained in it in a latent state, so as not to be sensible to the thermometer or to the eye; and if light be only a modification of heat, or a circumstance attending it, or a component part of the inflammable air, then pure or dephlogisticated air is composed of water deprived of its phlogiston and united to elementary heat?[68]

This section, itself italicized by Watt, is usually the only part of the paper quoted in discussions of the discovery of the composition of water. From it the nineteenth-century or modern reader, given the later equation of dephlogisticated with oxygen and of phlogiston with hydrogen (inflammable air), might be forgiven for thinking that Watt had the composition of water right except for some excisable complication regarding latent or elementary heat. Reading the next page of Watt's paper, however, undermines such a confident interpretation, and is to enter a world in which elements and compounds have yet to become the conceptual basis of chemistry. Instead we find a world of 'principles', 'affinity' and of substances uniting to varying degrees. Thus, Watt explains:

It appears that dephlogisticated water, or, which may be a better name for the basis of water and air, the element you [De Luc] call humor, has a more powerful attraction for phlogiston than it has for latent heat, but that it cannot unite with it, at least not to the point of saturation, or to the total expulsion of the heat, unless it be first made red-hot, or nearly so. The electric spark heats a portion of it red-hot, the attraction between the humor and the phlogiston takes place, and the heat which is let loose from the first portion heats a second, which operates in a like manner on the adjoining particles, and so continually until the whole is heated red-hot and decomposed.[69]

This reveals that Watt considered water and air to have a common basis. The language of 'bases' in this sense is like the language of 'principles'. These are the elementary materials of the world. So Watt is saying that water and air share an elementary principle. He gives this the name 'humor', a term applied in that sense by De Luc in his meteorological writings, which Watt was very familiar with.[70] The fact that Watt used this concept is a key indication that he saw fundamental unities beneath the diversity of water and airs, each being a variety of the other. The impetus for the reaction comes from the greater affinity that humor has for phlogiston than it has for latent heat. Of course, for Watt heat is a chemical substance and so has an affinity like any other. But red heat, created by the electric spark, provides the initial conditions under which the latent heat combined with a portion of the humor is converted into sensible heat. Once this occurs the humor unites with the phlogiston and the heat released initiates what we might now refer to as a chain reaction. Not everything was clear, however. Watt could not explain why the electric spark was required and why the reaction did not proceed at room temperature. He pointed to an interesting fact in this connection, namely that dephlogisticated air and phlogiston (inflammable air) 'can unite, in certain degrees' *without* changing into water. He cited an experiment by Priestley in which clean iron filings and mercury calx were heated together. Whilst heated separately they would produce inflammable air and dephlogisticated air respectively, when heated together they produced neither, but instead fixed air was found. From this, and from the fact of the long-term stability of a mixture of the two airs, Priestley concluded that heating the two substances together produced the airs in a *nascent* state, and in that state they could combine to produce not water but fixed air. So, depending on the circumstances, dephlogisticated air and inflammable air when in proximity could remain an inert mixture, could combine to produce water or could unite to form fixed air. Watt went further to suggest that other airs were also combinations of dephlogisticated air and phlogiston, phlogisticated air being a case in point. He felt he had grounds to believe that fixed air contained a greater quantity of phlogiston than fixed air did.[71]

It is apparent from all this, then, that Watt regarded the different airs *and* water as all being products of the uniting of dephlogisticated air and phlogiston in different degrees and by different means. This is the key insight that is missed when Watt's bald italicized statement on the composition of water is extracted and taken in isolation.

At this stage in the paper Watt recalled his prior beliefs about the relations of water and air and his reasons for them. First he had believed that air was a modification of water because 'in most cases wherein air was actually made ... the substances were such as were known to contain water as one of their constituent parts, yet no water was obtained in the processes except what was known to be only loosely connected with them, such as the water of crystallization of salts'.[72] Put another way, constituent water that was tightly connected disappeared during the making of air. Watt described this opinion as arising in its turn from a discovery about the latent and sensible heat of steam. He was referring here to the results of his steam experiments which he interpreted as showing that 'the latent heat of steam was less when it was produced under a greater pressure, or in a more dense state', the value varying in such a way that 'when a very great degree of heat was necessary for the production of the steam, the latent heat would be wholly changed into sensible heat' and the steam itself might undergo a remarkable change, becoming air. Watt's idea, then, was that tightly bound constituent water would require a high enough heat to release it so that all its latent heat would become sensible heat and the constituent water itself would change into air. The chemical significance of 'Watt's law' lay in this, so far as Watt's early beliefs about the mechanism through which water was transformed into air. Having explained these early beliefs, however, Watt said that he had now abandoned them because he thought that the transformation could be 'accounted for on better principles'.[73] Those better principles invoked the differential affinities of substances for humor and phlogiston:

> In every case, wherein dephlogisticated air has been produced, substances have been employed, some of whose constituent parts have a strong attraction for phlogiston, and, as it would appear, a stronger for that substance than humor has; they should, therefore, dephlogisticate the water or fixed air, and the humor thus set free should unite to the matter of fire and light and become pure air.[74]

Thus Watt still has water being converted into air, but this occurs not by the action of heat alone but by a process of dephlogistication of the water to produce the intermediate humor, which then becomes pure air by uniting with fire and light.

We can now understand why Watt spends the bulk of this paper dealing with experiments in which dephlogisticated air is produced in a variety of ways – he is attempting to show that air (in this case the dephlogisticated variety) is always produced from water via humor and not from some other constituent. In particular he believes that he can show that air is, as he put it in a letter to Priestley, 'not a child of acids', by showing that the products of the various reactions contain as much acid as was there at the beginning. Thus Watt recounts in detail his own and others' experiments with nitre, with magnesia alba, calcareous earth and minium (the red calx of lead), with mercury, sulphur and phosphorus. In all this he finds nothing to support, and much to deny, those views of the chemical origins of dephlogisticated air that trace it to sources other than constituent water, via humor.

Watt's second paper, immediately following the first in the *Transactions*, was a short one.[75] It gave advice about precautions to be taken by anyone repeating the experiments, and indicated some causes of variations in the results. Watt exhibited a typically firm diffidence about his own conclusions!

> The unavoidable little accidents which have attended these experiments, and which tend to render their results dubious have prevented me from relying on them as *full* proofs of the position that no acid enters into the composition of dephlogisticated air; though they give great possibility to the supposition. I have, therefore, explained the whole of the hypothesis and experiments with the diffidence which ought to accompany every attempt to account for the phenomena of nature on other principles than those which are commonly received by philosophers in general.[76]

Watt's contribution to the *Philosophical Transactions*, then, was not simply a bright speculative idea based only on Priestley's experimental work, but rather a sustained piece of experimental inquiry on Watt's own part. The long sections of Watt's Common Place Book[77] headed 'Dephlogisticated Air' recount his own experiments in even greater detail than is recorded in the *Transactions*. The entries begin on 3 May 1783, continuing through June and July and then resuming in November of that year. Priestley is a constant reference point and source of materials and ideas, but Watt remains his own man.

By July 1784 the Common Place Book is taking us beyond the views expressed in Watt's published papers. By then Watt was reacting to Lavoisier's interpretation of those phenomena. An entry for 13 July 1784 is headed 'Dephl^{d} Air, Mr Lavoisier from the Abbé Roziers Journal, Mai 1784', and recounts Lavoisier's 'imitation' of Priestley's experiments with the red-hot tobacco pipe. The imitation involved some variation, in that Lavoisier passed steam through a red-hot iron tube or dropped water slowly into one end of it, and then condensed any 'steams' that might pass through the red hot barrel. Watt put it thus:

> While the Iron is in this state the water acts upon it and separates the phlog^{n} or Inf^{ble} air, he says of the water, but more probably of the Iron, while the Dep^{d} air of the water as he supposes unites to the Iron & forms a calx such as has been described. It is more probable that the water in substance unites to the Iron & liberates its ph^{n} in ye form of Inf^{ble} air which is a comp[ositio]n of ph^{n} & ▽ [water].

Priestley later recalled the importance of Watt's insistence that Lavoisier could not interpret properly these experiments passing steam through a red-hot iron tube:

> I should not have done this [the repetition of Lavoisier's experiment] with so much advantage, if I had not had the assistance of Mr. Watt, who always thought that M. Lavoisier's experiments by no means favoured the conclusion that he drew from them. As to myself, I was a long time of opinion that his [Lavoisier's] conclusion was just, and that the inflammable air was really furnished by the water being decomposed in the process. But though I continued to be of this opinion for some time, the frequent repetition of the experiments, with the light which Mr. Watt's observations threw upon them, satisfied me at length that the inflammable air came from the charcoal, or the iron.[78]

This suggests, though it does not prove,[79] that Watt was instrumental in keeping Priestley true to the phlogistian faith at this juncture, and thus maintaining him on the course of opposition for the rest of his life to Lavoisier's claims about the decomposition of water.

The period from the mid- to the late 1780s was, of course, one in which Lavoisier's anti-phlogistic doctrine made great strides. Lavoisier's staging of the great experiment in 1785 in which water was analysed into its constituent gases that were then re-synthesized to form water, had a powerful persuasive effect. Lavoisier gained prominent supporters among French chemists and launched a more aggressive campaign in which his 1786 essay 'Réflexions sur le phlogistique', the *Méthode de Nomenclature Chimique* of 1787, and above all his 1789 text *Traité Élémentaire de Chemie* played a key part.[80]

Watt was more active behind the scenes in defending phlogiston against the French, and maintaining the resolve of its opponents than his sparse public pronouncements would indicate. We have seen that Watt was never far behind Priestley, and this continued to be the case. On 11 November 1784 Watt reported to Black on a series of 'capital Experiments' by Priestley and expressed the view that 'on the whole I think that poor Phlogiston will be perfectly saved from Mr. Lavoisier's sentence of annihilation, and I think myself bound in honour to help him up with his breeches, If I have Leisure'.[81] It was to be almost two years before Watt reported to Black again about new and interesting experiments by Priestley on 'disolving Charcoal by boiling in spirit of nitre that it produces a considerable quantity of air half of which is nitrous air and the remainder a mixture of fixt and phlogisticated air, but most fixt air…'. Watt also reported that Priestley had found some curious results by passing the 'steams of the nitrous and vitriolic acids over metals kept red-hot in an earthen tube impervious to air'. But Watt was in the doldrums himself, explaining to Black that 'my attention to philosophical subjects has much abated and I try no experiments of any kind myself'.[82]

Watt's chemical interests probably fell temporarily victim to a bout of low spirits. His experimental interests soon revived in a different, but related, context, that of his experiments on chlorine bleaching stimulated by his contact with Berthollet during the visit to Paris on engine business that Boulton and Watt made beginning in November 1786. Certainly by early 1788, by which time Priestley had conducted further experiments 'which bear hard on the doctrine of the formation of water from Deph^{d} and inf^{ble} airs', Watt was showing renewed interest. Though he expressed some doubts about the experiments, it appeared to Watt that 'a very large proportion of *all airs* even 9/10 is water, 2d that nitrous acid is a component part of D^{d} and inf^{ble} together form it and that probably phlo^{d} air is some modification of the other two'. Watt betrayed some frustration at the speed of claim and counter-claim between phlogistonians and their opponents – 'there it stands at present, but all may be changed again in a few months'. But he also detected in these experiments a 'fresh opening' for the phlogistonians.[83]

Watt remained prominent among their number and was acknowledged as such. Richard Kirwan was very active during this time. He gave an account of combustion, calcination and acid formation in which phlogiston released in combustion combined with dephlogisticated air to form fixed air, that air then combining with the residue of combustion to produce a calx or acid. As set out in *An Essay on Phlogiston and the Composition of Acids* (1787), this was an influential aspect of the rearguard action. We have seen above that Watt too adhered to the view that dephlogisticated air and phlogiston could and did combine in a variety of ways, not always forming water. Kirwan made use of Watt's explanation of how issues of heat content affected the viability of chemical processes. Writing about dephlogisticated and inflammable airs, Kirwan stated: 'both these airs may remain long together without forming any union, and when they do unite, it is because one of them has not its whole quantity of specific fire; but in high degrees of heat, their *specific* becomes sensible heat, as Mr. Watt has discovered'.[84] Priestley's ongoing writings also show us Watt still hovering in the background. Priestley's published account of the acid substances produced on the sparking of dephlogisticated and inflammable air incorporated letters from William Withering and James Keir, who assisted Priestley in his analyses. It also revealed Watt monitoring everything and offering suggestions: 'Mr. Watt desires me to mention it as his conjecture, that the nitrous acid is contained in the inflammable air as the acid of vitriol is in sulphur, the phosphoric in phosphorus, &c; and that the dephlogisticated air does nothing more than develop the acid'.[85]

Before moving to the next phase of Watt's chemical interests and development it is important to consider more generally the collaboration and relationship between Watt and Priestley. As W. H. Brock has recently explained, for modern chemists and traditional historians of chemistry Priestley is a highly problematic figure.[86] On the one hand he clearly deserves considerable credit for his work on the discovery of airs and is a candidate for the discovery of oxygen (along with Scheele and Lavoisier). Yet Priestley's apparently stubborn insistence on defending phlogiston theory beyond the point where most others had abandoned it has, for many, tarnished his reputation. He has also been portrayed as a slap-dash and unsystematic experimentalist, surprisingly and unforgivably, ignorant of chemical fundamentals, especially the chemistry of affinity and salts and acids within which Lavoisier was schooled. Even Robert Schofield, who has done more than anyone to rescue Priestley from his sternest critics, such as J. R. Partington, by resisting the worst of the Whiggish anachronistic judgements made about him, early painted Priestley's chemical career as on a downward trajectory from the time of his arrival in Birmingham, and thus from the time of his intimate acquaintance with Watt. Priestley, Schofield argued, was 'on the defensive' from that time onwards.[87] Watt, if this is true, was hitched to a fading star.

Whether or not Priestley, and therefore Watt, really was on the defensive through all or part of this period is a tricky question. To depict them as such requires a sense of the ascendency of others, and ultimately that sense can only be established retrospectively. Much historical exasperation with Priestley's work on airs stems from the assumption and feeling that he was, or ought to have been, working with a concept of 'gases'. But Priestley, like Watt, had no such conception. Although Watt's belief that all airs were modifications of a single air, or a smaller number of airs, almost certainly predates his close association with Priestley, that association undoubtedly reinforced Watt in that view. Neither man could escape the 'lumping' habit in their chemical thinking that the imperium of phlogiston and the chemical theory of latent heat imposed.

Turning from these broad historiographical questions to the more immediate experimental context, many people who have followed Priestley's experimental trajectory in the 1780s have found it easy to regard Watt as a speculator dependent upon the chaotic and undisciplined experimentation of his friend. Priestley's experiment with the earthenware retort in which he first claimed, and then retracted, the finding that water was being converted directly into air seems from this perspective to be a debacle. It led Watt to withdraw his letter to Priestley from the gaze of the Royal Society, and contributed to his loss of priority in the eyes of many. That loss was caused not just by delay but by the blow to Watt's credibility as a chemist that many subsequent interpreters found in that event. Typical of the line of thinking that sees Watt as suffering because he was a mere speculator is Richard Hills's remark that 'Perhaps Watt relied too much on Priestley's experiments at this time without carrying out many himself'.[88]

However, at the time that Priestley's experimental error was pointed out to him, Watt insisted that it did not destroy his theory about the nature of water. While Priestley ventured the observation that his discovery of the error 'utterly ruined your beautiful hypothesis',[89] Watt denied that that was the case:

> I maintain my hypothesis until it shall be shown that the water found after the explosion of pure and inflammable air had some other origin nor shall I believe that air is a child of acids or rather a modification of them untill such acids can be found after the decomp^{n} of it. I have many experiments to propose to you to help bring out the truth and a fair analysis of the two fluids.[90]

Many of these experiments, as we have seen, Watt conducted himself, recorded in his Common Place Book, and published in his 1784 *Philosophical Transactions* paper. They are our strongest evidence that Watt was not a mere speculator hanging from Priestley's fraying coat-tails, but rather a serious chemical thinker, not averse to devising and prosecuting his own experimental trajectory, and constantly learning from, and trying to make theoretical capital from, his practical projects.

We should not confine Watt's experimental experience to his activities in the laboratory. In a real sense his work devising, erecting and operating steam engines was also experimental and informed his chemical thinking. A good clue to this relates to Watt's response to Priestley's experiments in early 1783 'turning water into air'. Watt explained these developments to Joseph Black: 'water being put in a gun barrel and distilled over slowly gives no air but on being confined by a cock and lett out by puffs it produces much air, which agrees with my theory and also coincides with what I have observed in Steam Engines – In some cases I have seen the 10th of the bulk of the water of Air extricated or made from it'.[91] Watt had evidently shared these ideas with Priestley too since the latter remembered a conversation with his friend about the possibility of converting water into 'permanent air' in which Watt had observed that 'some appearances in the working of his fire engine had led him to expect this'.[92] How exactly does Watt's theory account for the difference between distilling water 'over slowly' and confining it and letting it out by 'puffs'? While he nowhere explains this fully, the putative explanation must be something to do with the longer retention of the water, which allows for more latent heat to be converted into sensible heat. We might ask also what was the analogous phenomenon that Watt saw in engines? Presumably it was the production of air from water: the last sentence of the above quotation must mean that he has seen air produced from water, either extricated or made from it, in steam engines, the resulting air being $1/10^{th}$ of the bulk of the water involved.[93] Whatever the precise explanation may be, the fact that Watt finds the same phenomena exhibited by Priestley's water experiments and by his steam engines reinforces my argument that Watt's chemistry and his steam engine work advanced together and were mutually informative in hitherto unappreciated ways. Watt's lifelong preoccupation with steam and steam engines continued to inform, and be informed by, his chemical thinking. But other practical projects in the late 1780s and 1790s stimulated Watt's chemical work also. It is to these that I now turn.

Watt's Chemistry in the late 1780s and Beyond

As already noted, in mid-November 1786, Watt and his partner Matthew Boulton travelled together to Paris. They had been invited to advise on improvements to the pumping station that supplied water to the Palace of Versailles, and they also hoped, vainly as it turned out, to finally clinch an exclusive privilege on their steam engines in France. There was time and opportunity to meet many of the leading natural philosophers of the city, who gave Watt in particular a rousing reception: he described himself as being 'drunk from morning till night with Burgundy and undeserved praise'.[94] Whilst the meetings with the likes of Lavoisier, Laplace, Monge, De Prony, Hassenfratz, Fourcroy and Delessert were exciting and tantalising given the ongoing water question, in many ways Watt most enjoyed his meetings with C. L. Berthollet. Those meetings and their correspondence over ensuing years were important, as historians have noted, in the development of chlorine bleaching in Britain.[95] Berthollet described to Watt the elements of that process and advised him about it as Watt tried to implement the new technology. What has not been appreciated, however, is the extent to which chlorine bleaching, and other exchanges with Berthollet, stimulated and revitalized Watt's chemical thinking more generally. It was through Berthollet that Watt communicated his response to the new system of chemical nomenclature being proposed by the French. Although the existence of Watt's response to that nomenclature has been known for some time, the extent to which it reveals his continuing adaptation but resistance to Lavoisian chemistry has not been fully appreciated. It is also likely that Watt's ideas about airs were shaped in important ways by Berthollet's affinity theory. This becomes apparent when we examine closely Watt's thinking about airs as it is revealed in the other great practical project of this period, his collaboration with Thomas Beddoes in the development of pneumatic medicine. A careful examination of that collaboration indicates that Watt, traditionally cast in the role of the engineer who devised the apparatus to deliver the supposedly curative airs, actually brought a good deal of chemistry to the venture and continued to build his chemical ideas through it.

Perhaps the most exciting event of Watt's time in Paris was Berthollet's demonstration of his method for bleaching cloth with dephlogisticated muriatic acid gas, that we call chlorine. Watt took careful notes of the process and immediately wrote an account of it to his wife Annie whose father, James McGrigor, was in the bleaching business. When Watt returned home he began experiments, which are carefully recorded in his Common Place Book, and instructed his father-in-law in trials of chlorine bleaching. Watt's experiments were designed to find the best method of application but also had a close eye on the economics of the process, upon which, in the end, everything would depend. By May 1787 he had a method that was ready to be scaled up. Hills comments, justly, on how impressive Watt's systematic, scientific approach to this problem was. The new method of chlorine production that he devised was capable of being used successfully, and safely, by inexperienced operators. Although he had obtained permission from Berthollet to use his ideas, Watt proceeded with great caution and in secret in order not to betray too much to others who were pursuing the same objective.

Even as he pursued these practical objectives Watt was still preoccupied with larger chemical questions. Indeed dephlogisticated muriatic acid air could not fail to stimulate in Watt thoughts about dephlogisticated airs in general, on which he had done so much work before, and which lay at the heart of key issues in chemistry. Probably through Berthollet and Hassenfratz, Watt became concerned with the new chemical nomenclature being developed in France. This led him to write a substantial chemical essay on the subject, first discussed in detail by David Larder.[96]

As the Lavoisian system gradually made converts in Britain, Watt was among those who maintained a position between the old and the new chemistry. Watt's response to the new chemical nomenclature of 1787 in his unpublished essay reveals his stance. His reaction to the symbolism devised by Hassenfratz and Adet is especially revealing. Whilst accepting its general value, Watt suggested a number of changes. Many of these were pragmatic ones, aiming for greater ease of use whilst working in the laboratory. Watt discussed many of these proposed changes with Berthollet to whom the essay was sent. [97] However, other suggested changes had significant chemical import. Watt wanted to extend the symbolism to emphasize the physical states of compounds. This was consistent with his longstanding belief that *chemical* processes account for changes of state.[98] Watt suggested changes in, or restated the meaning of, a number of the Frenchmen's symbols. Thus '|', for Caloric, he defined as follows:

> Caloric. Latent heat or matter of heat meaning that heat which enters into combination with substances & contributes to their assuming new forms, & which is not shown by the thermometer except when a substance changes its state.[99]

Under the symbol '..', Watt delineated his understanding of the term 'gas' as 'that modification of matter . . . not condensable by cold alone, which state seems to be produced by the union of heat & water & perhaps light with the respective bases, in some manner not yet perfectly explained'. When we take this statement in conjunction with his remarks on the gas 'oxygene' (symbol '—'), then Watt's conception of air becomes clear. Watt stated that Oxygene was the basis, or base, of dephlogisticated air. Thus for Watt at this point, dephlogisticated air was oxygene in combination with heat, water and perhaps light.[100] This represents a change in his thinking to the extent that he now seems to have abandoned the term and concept of 'humor', which earlier he had adopted from De Luc's writings. Now he takes 'oxygene' on board but not exactly as the Frenchmen would like. For Watt it remains a 'basis' of other things and its combination with heat and water is still chemically crucial. Watt depicted airs in like manner in his contributions to Thomas Beddoes's publications on the medicinal uses of airs, in the mid-1790s.[101] The tale of pneumatic medicine reveals another practical project in which Watt was involved, a project through which we can learn about the otherwise obscure changes and continuities in his thinking about airs in the 1790s.

In some respects the collaboration between Thomas Beddoes and James Watt in the study, promotion and development of pneumatic medicine in the 1790s is well known. The main public manifestation of that collaboration was their joint publication of *Considerations on the Medicinal Use and Production of Factitious Airs*.[102] The *Considerations*, in its very title and form, depicted the venture as a collaboration between Beddoes the medical man and Watt the engineer, Watt's contribution being presented as the design of an apparatus for producing, collecting and administering airs of possible therapeutic value. The company of Boulton & Watt produced and sold this apparatus. However, it is also well known that Watt had compelling personal reasons for his involvement, most notably the tubercular illnesses of his daughter Janet (Jessy) and his son Gregory. Jessy died before the planned Pneumatic Institute and its optimistic projectors had a chance to save her. Gregory, who died in 1804, survived long enough to know that Beddoes had abandoned hope of any major or lasting therapeutic outcome from the collaboration with his father. But there are signs that Watt retained hope in pneumatic medicine to the end of his life.

Most accounts of the Watt/Beddoes collaboration have purveyed a standard story, told with variations of emphasis.[103] That story has Watt driven by personal and humanitarian interest in finding better therapies for consumption. It portrays him as a politically and philosophically calming influence upon the more radical, volatile and enthusiastic Beddoes. It concentrates on Watt's construction of an apparatus, his hints and cautions about its use for production of airs and their therapeutic delivery to patients. There has been, however, no serious attempt to understand and interpret the ideas about airs that Watt brought to, and took from, the project. Nor have Watt's ideas been compared with those of Beddoes. One reason for the lack of attention to such questions may well have been the insistence of the protagonists themselves that theory had no place in their fact-based enterprise. They have been taken literally by most historians.[104]

Beddoes and Watt shared a chemical mentor – Joseph Black. Beddoes studied chemistry with Black whilst a medical student in Edinburgh in the early to mid-1780s, and when Beddoes began to lecture in chemistry at Oxford, his approach bore Black's imprint. He sought Black's advice about laboratory technique, and adopted the arrangement of chemical substances wholesale from Black's lectures. He told Black that he was equipped to demonstrate all Lavoisier's key experiments, and that he adopted a version of Lavoisier's chemistry in his own lectures.[105] Though in many respects well-equipped, Beddoes did not manage to mimic Black's success as a lecturer. His main problem was a lack of skill in experimental demonstrations.[106]

In late 1787 when he moved to Oxford to lecture on chemistry Beddoes seemed prepared to reject phlogiston and therefore the idea of a class of inflammable bodies that had it as a common principle.[107] However in the following year he had changed his mind again (along, it must be said, with a number of others), thanks to Priestley: 'Dr Priestley seems totally to have overthrown the antiphlogistic theory – I am anxious to hear what the French chemists have to say on the other side – I have seen some of their private objections to Dr Priestley's inferences, but they are totally insignificant'.[108] By 1791, however, Beddoes was once again convinced that Priestley and the phlogistonists were wrong. Beddoes only partially accepted the French chemistry, rejecting, for example, the notion of caloric as material heat.[109] He had long been an adherent of the 'new chemistry' so far as the understanding of airs was concerned by the time of his collaboration with Watt. When, in 1793, Watt shared with Black the news of Beddoes's Bristol practice he was in no doubt as to the chemical foundation of the enterprise: 'Doctor Beddoes is applying the antiphlogistic Chemistry to Medicine, Azote & other poisonous airs to Cure Consumptions & oxigene for spasmodic asthmas he is at Bristol wells for the greater practice'.[110]

Beddoes was an avowed Democrat in the years of the French Revolution and its aftermath, and while at Oxford had published pamphlets sympathizing with the revolutionaries, and expressing hostility to French clergy who sought refuge in England; such opinions attracted the notice of the Home Office, and were unwelcome in the Tory establishment in the university.[111] Thereafter Beddoes decided to pursue a medical, rather than a purely chemical, career.

Watt was first recruited to Beddoes's pneumatic medicine project in 1794 at about the time of his daughter Jessy's death from tuberculosis on 6 June that year. Richard Hills's account of the pneumatic project, which is probably based on a fuller acquaintance with the Watt correspondence than any other,[112] emphasizes Watt's personal grief as the chief driver of his interest and activity. Also, Hills is in little doubt that Beddoes used Watt to gain publicity and respectability. He suggests that Beddoes 'took advantage of Watt's bereavement and was quick to exploit the potential of his connections'.[113] In any case Watt quickly designed a simplified apparatus for producing various airs and delivering them to patients, and drew up a description of the apparatus with instructions for its use, which was published as Part II of *Considerations*. Boulton & Watt agreed to manufacture and sell the apparatus to Beddoes, to interested physicians and to members of the public. Between June and October 1794 Watt conducted numerous experiments and was in very regular correspondence with Beddoes (and a number of other physicians) about his findings and about other pneumatic business such as cases of treatment. He agreed, albeit reluctantly, to some of his letters containing 'hints' on the medical use of airs also being included in the *Considerations*.

Thereafter Watt's involvement became less intense, but he continued, occasionally, to experiment, correspond and advise. He cooperated in the production of new parts and editions of *Considerations* and also assisted in raising support for the Pneumatic Institute.[114] He reputedly installed the apparatus in the Institute when it was established in Bristol in 1797, and enjoyed cordial communications with the young assistant hired by Beddoes, one Humphry Davy. At the time of Gregory Watt's final illness in 1804, Beddoes and the establishment at Clifton still featured in Watt's desperate attempts to save his son, but whether he was treated with airs is unclear.[115] Beddoes's doubts about the pneumatic project were mounting by then. Thus, Watt wrote to Beddoes in 1803: 'You also seem to be among the sceptics in pneumatic medicine, you seem to have discarded both H. carbonate & oxygene which have done good here.'[116] Whilst it is unclear how long Watt held out serious hopes for the enterprise, we do know that as late as 1818 he retained enough belief in pneumatic medicine to take steps to try to make it available to his former employee and friend James Lawson. However, Lawson died before the apparatus that Watt had ordered to be sent to treat him could be supplied.[117]

In order to learn what light Watt's involvement in the pneumatic project can throw on his chemical ideas, especially his understanding of airs we need to examine in some detail what he, and Beddoes, have to say in the *Considerations*.

Beddoes's contribution to the *Considerations*, Part I, was more directly concerned with the physiological and medical aspects of the pneumatic project. Under normal atmospheric conditions and bodily functioning, he states, 'oxygene' air produces just the right degree of 'irritability' in the organs and other systems of the body. Too much or too little irritability will involve a diseased state of the body.[118] The taking of airs, particularly to vary the amount of 'oxygene' air taken into the blood via the lungs, is premised upon the idea of restoring just the right degree of irritability. Beddoes recounts his own, and others', experiments on animals, conducted to determine the relationship between the respiration of various airs and bodily functioning. The balance of Part I is devoted to cases reported to Beddoes by various practitioners. Importantly these cases are divided into four sections: first those referring to the administering of 'oxygene' air; second, cases involving administration of unrespirable airs; third, cases involving substances 'that may be supposed to act chemically'; fourth, 'miscellaneous facts and reflections'. Watt's hints are interposed between the second and third types of cases. This, we suspect, is for a very good reason, namely because Watt's focus is very much upon the way in which different airs deliver substances that act chemically. So his hints make the ideal transition between cases in which the focus is upon the chemical means whereby oxygenation is manipulated and those in which other chemically-acting substances are delivered to the lungs.

A significant difference between Beddoes and Watt was that Beddoes's main focus was upon airs as 'heightened' or 'lowered' whereas Watt concentrated upon the other chemical effects. This squares with the state of their broader chemical allegiances. Beddoes, as a convert to Lavoisier's oxygen chemistry (though not an uncritical one) is no longer preoccupied with the mysteries of airs, especially inflammable airs, in the way that Watt is. In his earlier publications, in which he first launched his interest in pneumatic medicine, Beddoes was very 'oxygen-centred'.[119] When Beddoes tabulated (in a 1796 publication) the cases he knew of in which factitious airs had been employed, the vast majority of treatments involved the breathing of 'oxygene' air in various dilutions.[120] Beddoes adopted the French nomenclature and so made it clear that his plans for studying and curing diseases by breathing of different airs were conceived within that system. He made the important point that most treatments would involve only small variations (from normal atmospheric composition as measured by Lavoisier) in the proportion of oxygen in the airs breathed.[121] So it was as a convert to the Lavoisian system, and with a focus very much on oxygen, that Beddoes launched his pneumatic medicine venture.

Watt's preoccupations and foci were rather different. As we have seen, he sought to hybridize the old and the new chemistry. He still believed in the interconvertibility of airs. Watt's preoccupation with the nature and role of charcoal is closely connected with his belief that charcoal is akin to (or to be identified with) phlogiston. As we will see, at one point Watt suggests that 'hydrogene' is nothing more than charcoal 'perfectly unified' with water. Watt was still puzzling over the nature of airs in a way that Beddoes was not, and this fact made the pneumatic project, at least so far as the chemistry of airs was concerned, a more complex, frustrating, but also more open-ended and potentially instructive, one for him than it was for his collaborator.

Watt contributed to the *Considerations* in two ways: in his account of the construction and use of the apparatus for producing airs and delivering them to the patient (Part II of *Considerations*), and secondly in a sequence of four letters written to Beddoes between June and October 1794 and included in the *Considerations* as 'Mr Watt's hints on the operation of different airs'.[122] With his habitual diffidence, in his 'hints' Watt apologized for troubling Beddoes with his 'crude ideas' on the use of pneumatic medicines and pleaded only the desire to help others avoid the parental sorrows that he had suffered as the reason why he had ventured 'to step over the bounds of my profession'.[123] In 'Apparatus' Watt's modesty took the form of an empirical repertoire advising that he had 'indifferently made use of the terms of the old and new Chemical Nomenclature, wishing merely to be understood, and not intending to enter into discussions upon theories in a treatise, the objects of which are facts'.[124] Despite Watt's announced determination to stick to the facts, we find numerous occasions in these works where we gain clues to the medical and chemical theories that informed Watt's thinking on pneumatic airs and their uses.

In some ways, Watt seemed to acknowledge airs as chemically distinct, notably when he discussed dephlogisticated, or 'oxygene', air. But there remain strong hints that he still held to the idea that different airs were modifications of each other. Thus, in 'Watt's hints' he states: 'remedies for some fatal or dangerous disorders may, possibly at least, be found in the class of airs, which admit of many *known* modifications, and doubtless many more to be discovered'.[125] The key issue here is what Watt means when he writes about 'modifications' of airs. Both Partington and Dyck have argued that Watt was exhibiting his ongoing adherence to his earlier view that different airs were the product of different combinations of heat with common air; that dephlogisticated, fixed and inflammable airs were in effect chemical varieties of the one air.[126] The difference between modifications of air and different species of air is important; even Priestley, the foremost pneumatic chemist of the day, was concerned with modifications of air, blurring the lines between species and varieties.

Something like that view holds most clearly in Watt's account of 'inflammable or hydrogene airs'. He distinguished these airs according to the mode of production. Inflammable air produced from zinc he described as the 'purest, or at least the lightest species of this air', whilst that produced from iron 'is the next in specific gravity to the inflammable air from zinc'.[127] From charcoal, water and heat, 'heavy inflammable air' is produced, also called by Watt 'carbonated Hydrogene, or Hydro-Carbonate'.

Watt, however, also appears to have had other phenomena in mind when he wrote of 'modifications' of airs. Both the practical production of airs and Watt's therapeutic hopes for them involved him in interesting ideas about the airs and the metals, or other substances, from which they are produced. These ideas were presented as a response to suggestions by 'some eminent physicians' that powdered substances, such as Peruvian bark and the calces of lead and zinc, might have therapeutic value in treating conditions of the lungs:

> To the use of powders, however finely *mechanically* divided, I think there are some objections; particularly I doubt whether they could enter the minute vesicles of the lungs, but if such substances can be *chemically* divided and obtained in the state of solution in air of some congenial species, they might have their full effect.[128]

This idea of 'chemical division' seems to imply an association between an air and a substance that is intermediate between mechanically separable and chemically compounded. The idea appears repeatedly in both 'Watt's hints' and in 'Apparatus', but is not expressed consistently. On occasion Watt uses the term 'suspension' to refer to it. For example, inflammable air produced from iron and vitriolic acid, he states 'always carries with it, even through water, a large quantity of iron; some of which it afterwards deposits, *but very probably some part still remains suspended*'.[129] The part of the carried-over iron that remains suspended must have a different relationship with the air than that which is deposited. That relationship is (from a modern perspective) an intermediate one between mechanical mix and chemical compound. For Watt this was simply a different form of compounding. Elsewhere Watt expressed this idea differently. Thus when referring to zincic inflammable air he distinguished between the suspension and the solution of flowers of zinc in the air:

> This air carries with it a large quantity of the flowers of zinc in suspension, which it deposits by standing at rest; it probably also contains another quantity in a state of solution, which seems to form a part of its substance, and on which some of its virtues may depend.[130]

The description of the zinc forming part of the 'substance' of the air by virtue of being dissolved in it is a revealing idea. It implies that there is no clear distinction to be made between solutions and chemical compounds – that chemical union admits of degrees of intimacy.

Watt's account of heavy inflammable air reinforced this notion. This air is produced, he says, by applying water to red-hot charcoal in closed vessels. It contains 'inflammable air, properly so called, fixed air, separable by water or by alkalis, and some other substance, which, when the inflammable air is deflagrated with oxygene air, produces fixed air. This substance I consider as charcoal in a state of solution'.[131] Watt took the fact that this 'charcoal in a state of solution' could not be removed in the same way as the fixed air (by passing it through water or alkali) as evidence that it was *not* 'fixed air completely formed'. And since he regarded fixed air as a saturated solution of charcoal in 'oxygene' air, the implication was that the 'charcoal in a state of solution' is an *un*saturated solution. In these ways charcoal was present in large quantities in heavy inflammable air. In fact, Watt said, 'no substance is dissolved in inflammable air in such quantities as charcoal, nor more intimately united'.[132] So, when Watt wrote about charcoal, or other substances, being dissolved in airs he was describing what he considered to be a process of chemical union that could exhibit degrees of intimacy.

This idea was very important to Watt's estimate of therapeutic possibilities. The anticipated efficacy of heavy inflammable air, and his special interest in it, was based in Watt's thinking on the amount of charcoal that it contained. Given the value of charcoal itself in 'correcting putridity', an air containing it should exhibit similar effects upon the body. However, in the case of treating consumption with airs, not just any old state of charcoal would do, according to Watt. The charcoal had to enter the bloodstream. Watt considered this unlikely to occur in the case of fixed air (a saturated solution of charcoal in 'oxygene' air) because the lungs would not be able to decompose it.[133] But the 'other substance' in heavy inflammable air that was an unsaturated solution of charcoal was more promising: 'Whether charcoal in this state could be decomposed by any excess of oxygene in the blood of consumptive patients, I cannot say; but it seems likely that it would, and at any rate it would act as charcoal powder does, and therefore highly merits trial'.[134] Overall it appears that Watt's chemistry of airs, as he mobilized it in the mid-1790s was still grounded in his early ideas of the interconvertibility of airs as modified by the subsequent debates on phlogiston theory, but that new ideas about airs as solutions had also been added. It is worth considering where those new ideas might have come from.

Watt's views on degrees of intimacy of chemical union and on processes of solution as creating compounds are reminiscent of ideas expressed by Claude-Louis Berthollet in his major works on affinity.[135] At first sight, however, the idea that Watt derived these concepts from Berthollet seems implausible because the Frenchman's major publications on the subject did not appear until the early nineteenth century. The possibility remains, however, because Berthollet's ideas on solution and indefinite proportions were present, albeit in unsystematic form, in his earlier publications, including his writings on bleaching and dyeing, with which, as we have seen, Watt was very familiar.[136] The intensive interactions between Watt and Berthollet make it very likely that the Frenchman's ideas on affinities and degrees of intimacy of chemical union, as expressed in his published papers and in his book on dyeing, were known to Watt who then deployed similar notions in thinking about airs when he turned to the pneumatic project. When Lavoisier in 1789 touched on affinities as the transcendental part of chemistry, he tipped his hat to Berthollet, whose researches in this field were already underway.[137]

Once we realize that Watt was working with this concept of degrees of intimacy of union, his ideas about the relative therapeutic value of airs become clearer. One clear statement of this is made in a letter that Watt intended to send to Dr Thomas Percival in November 1794. Writing of fixed air, Watt stated that it 'not only acts by exclusion of the oxygene, but also acts upon the nerves & upon the blood' and that fixed air thus has 'positive powers'. His guess from Priestley's experiments was that the fixed air would 'carbonize the blood'. By contrast, inflammable air (meaning here light inflammable air or 'hydrogene') 'seems to act principally [*sic*] by the exclusion of oxygene, unless when mixed with some other substance, such as those which give it smell, or the Iron or Zinc it contains'. Hydro carbonate, produced by the action of charcoal on steam, like fixed air, produced by the action of charcoal on 'oxygene', 'seems either to be rapidly absorbed by the blood or to act powerfully on the nerves'. Hydro carbonate 'seems to be the essence of fixt air'.[138]

Whilst Watt acknowledged those aspects of pneumatic medicine that involved the administration of excess oxygen or its reduction by processes of exclusion, he was particularly interested in what he called the 'positive powers' of airs that delivered substances to the blood and acted upon it. The ability to deliver depended upon how close the union of the substance in the air was. There appears to have been a hierarchy of effectiveness. Light inflammable air or 'hydrogene' air was the least effective in this regard because it consisted of 'the most perfect union of charcoal with water'. Perhaps the perfection of this union meant that the charcoal could not be given up to the blood whereas in the case of less perfect unions of charcoal in fixed air and hydro carbonate it could be so given up. This appears to be what Watt was trying to say to Beddoes in a letter of March 1795:

> I have formed an opinion, that, Oxygene & Hydrocarbonate are the only airs necessary to be employed. I think that the efficacy of Hydrogen (so called) depends intirely upon the Hydro carbonate it contains & that in fact it [the referent here is 'hydrogen proper'] may turn out to be nothing else than a more perfect union of charcoal with water... As to oxygen I am of opinion that in all ill conditioned ulcers, cancers not excepted that the breathing it will prove useful.[139]

An even fuller understanding of Watt's perspective on pneumatic medicine and the chemistry of airs can be gained by examining his correspondence with Beddoes and others at the height of his activity. The *Considerations* was first published in October 1794, some four months after Jessy's death on 6 June. In the intervening period, Watt (significantly aided by his younger son, Gregory) had engaged in a fevered bout of experimentation. Stimulated by the need to bury his grief in activity, and prompted by the hope of saving others from the fate of his daughter, Watt tackled many aspects of the production and medical use of airs during this period. This initial burst of activity was quite brief, and ended as rapidly as it began. Writing to Beddoes on 25 October, Watt observed: 'Gregory went off last night, & my laboratory is full of potatoes, so no more airs just now'.[140]

Although Watt did have some early involvement in the actual treatment of patients he found this uncomfortable and opted for self-experimentation. Watt dosed himself and reported the effects. On one occasion Gregory smelt an air and woke up prostrate staring at the ceiling. Watt and Gregory did most of their work on experiments that allowed Watt to provide hints to Beddoes and others about how best to produce different kinds of airs. While some products carried over with airs might be therapeutic, others, such as acids, could do damage and had to be removed. Watt also did much work on removing the smells of air, this not only to make them more palatable in treatment, but also to investigate whether the nausea induced by smell was the causal agent in any curative effect, or not. Beyond all this, though, Watt saw these experiments as part of an investigation into the nature of airs. His correspondence on medical and production aspects often switches to consideration of the chemistry of airs *per se*. Despite his disclaimer in the *Considerations* that in this work he was neutral between old and new chemical systems and was just offering facts, the facts that Watt produced he saw and valued precisely as they threw light on the phlogiston issue.

Describing to Darwin the apparatus that he had developed, and how it would be used, Watt reminded his friend 'I am still a sceptic as to the antiphlogistic theory in general though I believe some part of it'.[141] Writing to another correspondent a few months later, outlining his approach to pneumatic medicine and the facts on which his views were based, Watt explained his habitual use of the term 'inflammable air' rather than 'hydrogene': 'I dislike the word Hydrogene as possibly not founded in fact'.[142] In these little ways Watt reminded his correspondents that his approach to pneumatic medicine was, for all his vaunted even-handedness about facts in public, based on a theoretical position at odds with the new chemistry.

The clearest evidence for Watt's theoretical views on airs at this time, but in a frustratingly fragmented form, comes in correspondence with Beddoes just as they were finalizing the proofs of *Considerations*. In a letter of 3 September Watt mentioned that he had enclosed a '2d memoir upon airs'. This memoir has not been located, but Watt's previous letter to Beddoes (itself imperfectly preserved) contains a disquisition on airs that may well have been a first memoir. The status of these memoirs is unclear, but they seem to have been separate from the material that Watt contributed to the *Considerations*. That material was referred to in correspondence as the 'description of the apparatus' or as 'hints' on airs, not as a memoir or memoirs. Perhaps the memoirs were contemplated for inclusion in the *Considerations*, although their explicitly controversial character would have been completely out of tune with the rhetorical stance otherwise adopted in that publication. Maybe Watt was considering publishing them elsewhere. Whatever the case, the fragment of what is probably the first memoir on airs refers to 'another proof that no air can be made without water & that it forms the principal part of its gravitating matter', and continues, 'Airs, then, consist of water or steam united to some other substances, all in themselves unknown, except charcoal'.[143] A few days later Watt added another comment that clarifies his point:

> My opinion I have hinted before that water in specie forms the greater part of all airs, & I think the heavy inflammable air should be named Hydro carbonate for if it can be produced in so small a mixture as 1/12 of fixt air as Dr Priestley affirms what becomes of the oxygen if the water is supposed to be decomposed[?][144]

Hydrocarbonate is so important to Watt because it is the archetypical air produced by dripping water onto red-hot charcoal. If the water was decomposed in the process then one would expect a significant proportion of fixed air to be produced. Since this was not always the case, the water must not be decomposed but rather enter into airs *in specie*, that is, in the actual form of water.

Watt here is making it clear that he is aligned with Priestley in fighting what Darwin, Beddoes and many other British chemists by this time regarded as a rearguard action against the 'new chemistry'. Watt and Priestley had, as we have seen, long worked closely together and, although their relationship was not so close in the 1790s, they continued to refer to each other's work to bolster their position. After his departure from Birmingham after rioters attacked and burned his house, Priestley resumed work in London and from there in 1793 published *Experiments on the Generation of Air from Water*. In this he reprinted a *Philosophical Transactions* paper in which he argued that it was probable 'that water, as I have lately advanced, is only the *basis* of those kinds of air [dephlogisticated and inflammable] as well as of every other'.[145] This work was dedicated to the members of the Lunar Society and anticipated Priestley's imminent departure for the United States. There, as is well known, Priestley continued to write in support of the phlogiston theory. He also took an interest in Watt's work on medicinal airs, while Watt sent him an apparatus.[146] In *The Doctrine of Phlogiston Established*, Priestley noted that his experiment showing that the slow passage of steam over red-hot charcoal produced nothing but inflammable air had been 'confirmed by Mr. Watt, whose accuracy no person will call in question':

> He [Watt] says in his *Description of a pneumatical apparatus*, subjoined to Dr Beddoes's *Considerations on the medicinal use of factitious air*, p. 84 'It has been observed by Dr. Priestley, and confirmed by my experience, that when much water passes in the

> form of steam, there is much fixed air formed; but little or none when the water is admitted so sparingly that no steam reaches the refrigeratory'.[147]

Priestley, if not Watt, was happy to implicate Watt's contribution to the *Considerations*, publicly as well as privately, in a rearguard defence of phlogiston theory.

For Watt, as we have seen, the theoretical background to his understanding of airs, and to his hopes for their successful therapeutic use, was more complex. Whilst he saw some virtues in the new chemistry, he remained a phlogistian. His theoretical reflexes in the 1790s still harked back to ideas formed early in his chemical investigations about the inter-convertibility of different airs. The puzzle for Watt was still very much framed as about finding out what it was that accounted for interconversions. But despite these deep continuities in his thought, Watt drew (in magpie-like fashion) on other chemical ideas. It appears that Watt adopted ideas about solutions and degrees of chemical union from Berthollet as a result of their exchanges about bleaching and dyeing, which he then deployed in his thinking about airs in the 1790s.

Conclusion

As Dyck pointed out, Watt's distinctive chemical explanation in his work on the composition of water was lost during the assimilation of Black's doctrines to the new Lavoisian system in the 1780s and 1790s. Watt had pursued some of the explanatory possibilities that Black's ideas offered for conceptualizing chemical change in terms of latent heat in a way that the great teacher himself had not. Once Black's ideas were subsumed into another system, Watt's work would come to seem idiosyncratic. Even by the late 1780s and 1790s this process was underway, with Black contributing to the process himself in his lectures by his incorporation of Lavoisian ideas. Occasionally something of the Scottish tradition in its unassimilated form could be seen, Robison's articles in the third edition (1797) of the *Encyclopaedia Britannica* being a prime example.[148]

My point about Watt's chemistry, then, is a simple but little-heeded one. To understand its character and substance we need to assess the way that Watt drew upon a series of collaborative relationships, a number of them connected directly with practical ventures. First, we should look to his involvement with the Black tradition and the resources available to him from predecessors such as Boerhaave, Cullen, Eller and the like. From these resources, and from his own experimental work on the properties of steam (in the laboratory and, it must not be forgotten, in the large through the actual construction and operation of engines), Watt developed ideas about water and steam in which they were compounded with heat. This re-contextualization of Watt's chemistry makes it clear that throughout his active work on the steam engine, and consistent with his larger chemical views, for Watt steam was a *compound* of water and latent heat. Watt found support for his ideas in the meteorology of De Luc and in his chemical collaboration with Priestley. That collaboration was based in part on Watt's predilection for the theoretical interpretation of the experiments of others, but this does not mean that Watt was entirely dependent on Priestley for experimental fodder, as many accounts have assumed. Watt conducted numerous experiments of his own, especially on dephlogisticated air. These, together with his earlier steam experiments (to which he continually returned with a view to refining and extending them) gave him an independent experimental base from which to develop and maintain his chemical ideas about water, steam and airs. Once we realize this then Watt is no longer quite the victim to the experimental failures of Priestley that he is usually portrayed as.

Watt's publications in the *Philosophical Transactions* of the Royal Society in 1784 were important in themselves, and also because they opened a wider world of chemical contacts to him. His visit to Paris in 1786 brought him within the ambit of a number of important French chemists. These included Lavoisier himself, but arguably the most important of them from Watt's perspective was Berthollet. For it was Berthollet who not only confided his methods of chlorine bleaching but also discussed with Watt, in their correspondence, larger questions of chemical nomenclature and process.

The accumulated knowledge and technique thus acquired from his multiple collaborations and his own extensive experimental work were brought to bear on his final major collaboration, with Beddoes on the pneumatic medicine project. I have argued that in this project, too, Watt's work – usually seen as the practical contribution of an engineer – was in fact both chemically informed and chemically significant. Watt certainly hoped for medical benefits from the venture but his chemical curiosity was also piqued. His emphasis on the potential medical value of hydrocarbonate was a result of his theoretical reasoning about the nature and mode of action of that 'air'. Three chemical components – the chemical theory of heat, 'oxygene' theory reconciled to the continued importance of phlogiston, and an affinity theory which allowed for the combination of substances in varying degrees – all entered into that reasoning. Insofar as treatments were found to work – and there were many hopeful signs – the pneumatic project provided further justification for the value of the assembly of chemical understandings that Watt brought to it.

That Watt's chemical understandings were later to represent a crime against progress for characters such as William Whewell, or an embarrassing archaicism for Watt's nineteenth-century supporters in the water controversy, should not be allowed to obscure them for us now. Getting Watt's chemistry in focus makes clearer its thematic importance to virtually all his practical ventures, including his steam engine innovations. It is to that specific connection that I now turn in the final two chapters.

5 THE STEAM ENGINE AS CHEMISTRY

There are two ways in which the centrality of the chemistry of heat, water and steam to Watt's steam engine improvements can be discerned. One, which will be pursued later in this chapter, is to reconstruct in a technical sense – building on what we now know about Watt's chemistry – the conceptual world in which, and through which, Watt's thinking about the steam engine took place. A second is to gain an appreciation of what we might call the 'ecology of steam'. By this I mean the broader chemical cosmology, embracing meteorology and geology, of which steam and the steam engine were a part, for Watt and his contemporaries. I start with the wider chemical picture.

The Ecology of Steam: Meteorology, Geology and *The Botanic Garden*

Newcomen engines are known as 'atmospheric' engines. This is because the drive on the piston is provided by the atmosphere working against the partial vacuum created in the cylinder by the condensation of the steam. But there was, and is, a deeper sense in which steam engines were atmospheric engines even when Watt put the steam itself to work in driving the piston. The engines were seen as trading on processes that were ubiquitous in the natural world and responsible for other atmospheric and geological phenomena. Meteorological and geological understandings were thus important to the science of heat and steam that underpinned the steam engine. Those who thought about steam and the steam engine sought, and found, coherences between phenomena in the realm of engines, meteorology and geology.[1] Indeed, I suggest that a steam engine was sometimes thought of as a kind of local corralling of the wild forces of nature. In the twentieth century nuclear physicists sometimes spoke about the act of deploying nuclear forces to provide energy via nuclear reactors as 'twisting the tail of the dragon'.[2] I believe that the same sense of domesticating a wild phenomenon attended eighteenth-century steam engine technology. The steam engine in the eighteenth-century landscape was, not just aesthetically, but also ecologically, part of what went on around it.[3]

Watt and De Luc

Watt had a longstanding relationship with the Swiss meteorologist Jean André De Luc. By the time De Luc began his extended residence in England as Reader to Queen Caroline in 1775 Watt was already familiar with his work, having reported to William Small in 1773 that he had 'been reading De Luc lately, and I have tried a curious experiment to determine the heats at which water boils at every inch of mercury from vacuo to air. De Luc's observations and mine agree; but his rule is false'.[4] The point to note here is that De Luc's writings on the atmosphere (in this case his *Recherches sur les Modifications de l'Atmosphere* published in 1772) appear to have inspired Watt to conduct particular chemical experiments. Moreover, although these experiments were most immediately connected with Watt's efforts to construct barometers, they were also part of his attempts to develop a deeper understanding of the steam engine.[5] Thus, immediately after reporting on the repetition of the De Luc experiment, Watt advised:

> I have some thoughts of writing a book, the 'Elements of the Theory of Steam-engines', in which, however, I shall only give the enunciation of the perfect engine. This book might do me and the scheme good. And would still leave the world in the dark as to the true construction of the engine. Something of this kind is necessary, as Smeaton is labouring hard at the subject; and if I can make no profit, I ought not to lose the honour of my experiments.[6]

It seems clear from the way that Watt juxtaposed his reading of, and views about, De Luc that he conceived a close relationship between De Luc's chemical meteorology and his own experiments on steam. Certainly he was to find fault with some of De Luc's ideas and was cautious about ensuring that De Luc did not overlook the credit due to Watt's friend and mentor Joseph Black in the discovery of 'latent heat'. However, their meteorological and steam engine interests were about atmospheres in a way that was mutually informative. It seems that in 1773 De Luc's writings were one of the factors that turned Watt to the thought of developing and writing his theory of steam-engines.

By the time De Luc published his *Idées sur la Météorologie* in 1787, Watt was not only familiar with the ideas expressed there but had learned about many of them firsthand and prior to publication. Watt's experiments in the early 1780s, which were so important to his ideas on both steam and the composition of water, were developed in concert with De Luc's work. In fact, when Watt wrote the famous letters to Matthew Boulton and to Joseph Black (discussed in Chapter 4), in which he announced his ideas about the composition of water he had been working directly with De Luc in the laboratory on steam, heat and evaporation. As we have seen, Watt's paper on the composition of water in the *Philosophical Transactions* in 1784 took the form of a letter to De Luc, and, of course, De Luc was the one who urged Watt to assert his rights in the controversy with Cavendish.

De Luc was in Birmingham in October 1782 and visited Soho in the company of Baron Reden.[7] Watt showed the visitors around and was impressed by De Luc as 'a modest ingeneous' man. This was probably their first meeting. It was either on that occasion or on a return visit on 9 November that De Luc told Watt about his plans to write 'some thing on heat and on the nature of elastic fluids'.[8] Watt reported that during the November visit De Luc came for breakfast and 'spent the whole forenoon incensing himself with steam & steam Engines'.[9] De Luc was anxious to learn about Watt's experiments, and Watt was forthcoming to some extent. The difficulty was that he could not explain his own work without referring to Black's on latent heat and he wanted to obtain the professor's permission before he gave too much information to De Luc.[10] It was at this juncture that Watt conveyed his idea about the composition of water to Matthew Boulton. This, it will be recalled, involved the idea that water was converted into air if heated enough. For Watt this also had meteorological import – 'here is a plain account of where atmospheric air comes from', he suggested to Boulton.[11]

After the late 1782 visit, De Luc was due to return to Birmingham, and it appears that he did return in March 1783, staying with Watt.[12] According to Watt, during that visit he and De Luc were 'principally employed in experiments of the latent heat in vacuo, but from the many causes of inaccuracy could come to no other conclusion than that at least the sum of the latent or sensible heat is always equal, that is that the latent heat encreases [*sic*] as the sensible heat decreases and in some experiments the sum of the heats appeared to be greater *in Vacuo* than in the Air'.[13] De Luc was at Watt's house again in late September 1783 conducting further experiments.

So Watt's engine work and De Luc's meteorology certainly had plenty of opportunity to intersect. In fact, substantially similar chemical ideas were deployed in each domain.[14] This intersection is apparent retrospectively in De Luc's *Idées sur la Météorologie*. A centrepiece of the *Idées* was De Luc's critique of the solution theory of evaporation as put forward by Senebier, Franklin, Hutton and others. According to that theory, water vapour was produced – that is, evaporation occurred – by the dissolution of water in air in a way precisely analogous to the dissolution of salt in water. This theory faced the difficulty that evaporation occurred in vacuo. De Luc's alternative theory was that water vapour was in fact a chemical product of water and heat. In recounting how he became confident of that alternative theory De Luc gave a prominent place to Watt's steam experiments.

In the *Idées* De Luc noted how Watt applied and extended the ideas of Joseph Black on latent heat. Watt applied them to the steam engine 'where the vapours of boiling water played such a large part'. Watt had 'already surmised that these machines were very far from producing the amount of effect of which their cause was capable' and, De Luc stated, 'he [Watt] attributed that deficiency to lack of knowledge about the nature of the active agent'. De Luc also recalled how Watt kindly repeated his experiments for him [De Luc], and the Swiss reported in detail those of Watt's experiments 'which helped me most directly to demonstrate that the latent heat of vapours cannot be explained except by a *chemical* union of particles of fire and those of water'. So De Luc stated that Watt's steam experiments provided him with the key evidence for the central proposition of his meteorological work.[15]

In tracing connections between steam engines and meteorology, it is also worth noting that De Luc, who was perhaps overly fond of analogies, invoked images of boilers in his discussions of clouds. Having argued that the appearance and disappearance of clouds was independent of the hydroscopic state of the air (and therefore not just a matter of solution and precipitation), De Luc went on:

> Observing these metamorphoses of one and the same cloud, nothing represents them more exactly for me than the idea of an invisible boiler suspended in the air, with water boiling violently in it. This image is too perfect a symbol of what is then observed for there not to be some analogy in the causes; that is to say, some metamorphosis of substance.[16]

Ultimately, De Luc's great 'discovery' was that evaporated water, which, if it accumulated in long periods of fine weather could not possibly all conceal itself in the upper atmosphere, 'hides itself in the atmosphere under the appearance of some gaseous fluid [fluide aëriforme]; for this is the only way in which it can escape its pressure, growing upon it as it accumulates, and the great differences of temperature to which it is subjected'.[17] Having thus been hidden, the gaseous fluid is on occasion converted back into its watery form as vapour, becoming once again available to form clouds and rain. Thus the interconversion of air and water was the basis of De Luc's theory of rain and that interconversion itself depended upon the composition and decomposition of water particles and fire particles.

The interconversion of air and water gained credibility when De Luc learned of the formation of water by the sparking of inflammable air in dephlogisticated air. De Luc recalled: 'the idea of water in any aërial form whatever was to me what a meeting of sea birds would be to sailors who have lost their compass and are searching blindly for the shore'.[18] We must remember that many regarded the various airs that had recently been discovered as different varieties of the one, common, air, so that any gaseous product of water would be regarded as 'air' and would establish the interconvertibility of the two. We have already seen[19] that Watt used De Luc's concept of 'humor' in his 1784 *Philosophical Transactions* paper on the nature of water. That concept too underscored the idea that water and air were extremes on a chemical spectrum. Both Watt and De Luc were eager to deal with the various 'forms' of water, including air, as variants upon the 'humor' principle.

Even more exciting, for both De Luc and for Watt, at least initially, was Priestley's announcement in 1783 of his conversion of water into air in an earthenware retort merely by heating. We have seen that this encouraged Watt to put forward his long-gestating theory of the composition of water, and confirmed De Luc in his theory of water vapour and rain. When Priestley subsequently reported that his experiment had been in error, both Watt and De Luc denied that this reversal put their theories in danger. Watt had merely denied that it demolished his theory. De Luc refused to believe Priestley's retraction.

Once again it must be recalled that Watt and De Luc were working very closely at this point, the one concerned primarily with his water and steam-engine work, the other with his meteorology. Just how closely Watt and De Luc converged in treating steam engines and meteorological processes as physically related remains unclear. Theodore Feldman, having examined their work and found it conceptually related up to a point, concluded, however, that 'Watt was experimenting with and for steam engines. He did not put his work in a meteorological context'.[20] Feldman argues that Watt in fact believed in the solution theory of evaporation so far as evaporation at normal temperatures was concerned. But Feldman grounds this claim on views apparently consistent with the solution theory of evaporation that Watt expressed in letters written in 1779 and 1780. These letters were written before Watt's personal encounters with De Luc, which clearly much impressed him. In late 1783 Watt wrote to Gilbert Hamilton:

> Mr De Luc is here now, he is busy writing a dissertation on spontaneous evaporation, which he contends is not caused by a solution in air but by Latent heat, in the same manner as vapour is produced by boiling water. I am giving him all the assistance in this subject I can, and he makes frequent and honourable mention of me in the work which is of consequence as it is to be published at Paris.[21]

Although divining Watt's meaning here is difficult in some respects,[22] this passage does suggest that he was not confident enough, or eager, to flatly reject De Luc's account of evaporation, at least in late 1783. It is possible, indeed, that at that time Watt agreed with De Luc. At the very least he was not worried about having his name and reputation associated with De Luc's ideas or about being seen as his collaborator. He was clearly flattered by it, and enjoying the prospect. It may well be that at this point he was prepared to accept De Luc's intended extension of the imperium of the chemistry of latent heat. What is certain is that for both of them a chemical theory of heat held the key. De Luc affirmed the chemical nature of meteorology when he envisaged the atmosphere as 'a chemical laboratory, as important as the bowels of the earth for the physical phenomena of our globe'.[23] Mention of the 'bowels of the earth' points us to another aspect of the 'cosmology' underwritten by the chemistry of heat that framed Watt's work.

Watt, Hutton and 'Geology'

It is worth considering briefly the role of the chemistry of heat in the development of geology during this same period, especially its role in the geology of Watt's friend James Hutton. Here too the steam engine and geological processes were analogically and conceptually interlinked in revealing ways. Also, as we saw in Chapter 4, it has been shown by Patsy Gerstner and Arthur Donovan that Hutton's ideas were subsequently 'cleaned up' by his followers in order to make them more acceptable in the rapidly changing natural philosophical landscape of the early nineteenth century. My claim is analogous: that a very similar thing happened to James Watt's ideas about steam and the steam engine in a way that effectively disguised their origins in the chemistry of steam.

Hutton's theory of heat was very important to his theory of the Earth but was given little prominence in the accounts of Hutton's theory developed by his followers, notably John Playfair and Sir James Hall,[24] who reinterpreted and reshaped it. For them the chemical theory of heat was an excisable irrelevance, a diversion from the centrally important features of Hutton's theory of the Earth.[25]

Hutton used the ideas of latent and specific heats, which were respectively the principles of fluidity and volume, as parts of the 'repulsive force' or what he called the 'solar substance'. Together with light and electricity, specific heat and volume made up the repulsive force. Hutton explained the dynamics of natural cycles of rock formations largely in these terms within his theory, and so Hutton's theory of the Earth was given its dynamics by the chemical theory of heat.

We have seen how De Luc pursued analogies between meteorological processes and those in the steam engine, and we will see below that Erasmus Darwin did the same in various of his writings. Hutton was also inclined to these analogical transpositions with regard to his theory of the Earth. A number of historians of geology have observed that Hutton was much impressed by the new steam engines which, with their complex, large-scale movements driven by a single source of heat, may well have been one of the inspirations for his theory of a heat-driven earth machine.[26] If this was the case then the steam engine was effectively treated as a microcosm of the earth-machine. These ideas were certainly prevalent in more popular accounts of the steam engine in the late eighteenth and very early nineteenth centuries. Thus the entry in *Public Characters* for Matthew Boulton, when giving an account of the steam engine, states: 'Thus we "put a hook in the nose of the leviathan"; thus we "play with him as a child and take him as a servant for ever"; thus "we subdue nature, and derive aid and comfort from the elements of earthquakes".[27]

When we trace the relations of Hutton and Watt we find that they had a close association for many years. When they met is unclear, but it was sometime before 1772 and probably through their mutual friend, and common chemical mentor, Joseph Black.[28] There were plans at one stage for the three of them to travel together on the occasion of Watt's historic removal to Birmingham in 1774, but in the event only Hutton left with Watt on 17 May that year. They arrived in Birmingham on 31 May. Watt and Hutton visited the Cheshire salt mines either during that journey or shortly after their arrival in Birmingham and in those mines Hutton made observations later to be important in his *Theory of the Earth*.[29] It seems possible that this period of close association and work together cemented what was to be a long, if physically distant, friendship. Their subsequent surviving correspondence is scanty but intimate. In his more regular correspondence with Black, Watt constantly referred to, inquired after, and asked to be remembered to, his friend Dr. Hutton.[30]

In Birmingham and its environs Hutton met other members of the Lunar Society, including Matthew Boulton and Erasmus Darwin. On a visit to the latter in Lichfield, Hutton used Darwin's house as a base for expeditions into Derbyshire and also participated in an experiment with an airgun and thermometer to demonstrate how the expansion of air cools it.[31] Darwin was later to make use of this fact in explaining processes of 'devaporation' in clouds and in steam engines, as we will see in a moment.

At the time of their journey together Hutton and Watt must have discussed pretty thoroughly their respective preoccupations. We know from Watt's correspondence, as already noted, that he had recently been conducting experiments on the boiling of water at various pressures, partly inspired by De Luc's work.[32] Given that Hutton was already preoccupied with the role of fire in the formation of the rocks of the Earth, and the behaviour of materials, including aeriform fluids, at high pressures, it seems likely that the similarities between their respective inquiries would have been discussed. Hutton was to rely on ideas about the way in which extreme heat and high pressure brought about departures from normal chemical action. For example, he explained the non-caustic character of calcareous earths – a property urged by some as evidence against Hutton's theory of heat as the primary agent – by the fact that high pressure had prevented heat from driving out the fixed air as it did under normal laboratory conditions.[33] Watt, as we have seen, developed the idea that extreme heat would convert water into air, not just into vapour.

A number of historians of geology have been persuaded that Hutton's exposure to Watt's steam engines probably provided Hutton with the idea of the Earth as a machine, which is a recurrent analogy in his theory of the Earth. Playfair's biography of Hutton noted the admiration that he expressed for Watt's engine,[34] and this observation has since been supported by other writers. Gordon Davies speculated that Hutton may have witnessed Watt's first full-scale experiments with the steam engine in 1769 at Kinneil House, the home of Watt's first patron John Roebuck, which was not far from Edinburgh. We know that Hutton visited Kinneil because he refers in the *Theory* to a bed of oyster shells seen there. This observation could have been at the same time as he witnessed Watt's experiments. It is even more likely that Hutton saw the Kinneil engine when it was re-erected in Birmingham for his fellow traveller Watt in 1774.[35]

Hutton's 'Earth-machine' might in theory have been inspired by any of the machines of the Industrial Revolution, but it is surely no coincidence that he regarded the Earth as a kind of heat engine, which brings the analogy closer. Whether the analogy was even closer still depends on whether Hutton can be reasonably attributed with the idea that the burning of coal was the source of the Earth's subterranean heat – an issue that divides historians of geology.[36] But even if coal was not the fuel of the heat machine, the analogy remains strong. As Gordon Davies notes, 'the active agents in Watt's engine and the Huttonian theory are identical – fire and water':

> Hutton himself observed that his theory is 'a system in which the subterranean power of fire, or heat, co-operates with the action of water upon the surface of the Earth…'. He perhaps saw the elevation of new continents by subterranean heat, and the lowering of the older continents by atmospheric agencies, as in some way analogous to the rise and fall of the piston in one of Watt's early single-acting pumping engines.[37]

If De Luc liked to think of steam boilers in the clouds, perhaps Hutton had pistons deep within the Earth! As Davies also notes, the Earth/heat-engine analogy is reinforced by Hutton's treatment of volcanoes as a kind of safety valve. Hutton put it this way:

> A volcano is not made on purpose to frighten superstitious people into fits of piety and devotion, nor to overwhelm devoted cities with destruction; a volcano should be considered as a spiracle to the subterranean furnace, in order to prevent the unnecessary elevation of land, and fatal effects of earthquakes.[38]

Hutton, by using the idea of a safety-valve in this way, deepens the analogy between the Earth-machine and the steam engine and reinforces our belief that thinking about steam engines and about the Earth had a great deal to do with each other for *both* Hutton and Watt. If I am right in treating Watt's steam engine as a chemical and mechanical device then it certainly squares well with Hutton's vision of 'this earth as a machine, constructed upon chemical as well as mechanical principles, by which its different parts are all adapted, in form, in quality, and in quantity to a certain end…'.[39]

There is some evidence that geological questions were likely to spring to Watt's mind when he conducted steam experiments and experiments on airs. This was certainly true in the early days of his collaboration with Priestley. Indeed, understanding Priestley's and Watt's chemistry involves appreciating the 'connectedness' that they saw and assumed between phenomena that we have divided and demarcated. Our own later maps of knowledge impel us to subsume every move that Priestley made under controversies about 'oxygen', 'phlogiston' and the 'composition of water'. This is a mistake. Robert Schofield long ago pointed out the danger of seeing Priestley as reactive (to Lavoisier and the theory of oxidation) when he was pursuing his own positive programmes of research:

> It is not immediately obvious ... that his opposition to the theory of oxidation was a compulsive reaction which was to dominate Priestley's research until his death. Indeed the first set of experiments [undertaken in Birmingham] seems wholly unrelated to that controversy...the immediate motive appears to have been the geological interests of his fellow members of the Lunar Society.[40]

Priestley published on this work in a note prefixed to a paper by William Withering in the *Philosophical Transactions* in 1782. Priestley was examining the extraction of air from various rocks and minerals, a line of inquiry that he reported on more fully in *Experiments and Observations* in 1786.[41] The quantity of air released on heating was taken to be an indication of the mode of formation of the rock. Priestley argued, for example, that basalt and granite when heated produced too much air to be the product of volcanic action. These experiments bore directly on the issues at stake between 'plutonists' who regarded fire as the primary geological agent and 'neptunists' who stressed the importance of water.

Schofield detects important continuities between these experiments and Priestley's earlier and later ones. All of them related to the natural economy of air – how air was produced in the first place, and its consumption, production and revitalization through various processes. Thus, in one of his first important chemical papers Priestley wrote:

> Considering this amazing consumption of air, by fires of all kinds, volcanos &c. it becomes a great object of philosophical inquiry, to ascertain what change is made in the constitution of the air by flame, and to discover what provision there is in nature for remedying the injury which the atmosphere receives by this means.[42]

When, in the early 1780s, Priestley experimented on the extraction of air from substances of volcanic origin he was still concerned to gain insights into the consumption of air by volcanoes. And when, in what are usually seen as key documents in the water controversy, Priestley heated water in contact with calcareous substances and claimed to produce air from water, he was completing the circle by showing how atmospheric air may have been produced in the first place from water, and how natural processes might remedy its consumption. The experiments of John Warltire sparking mixtures of inflammable and ordinary airs to produce water promised further insights into the same natural economy.

Watt shared the broad outline and many of the details of this coherent conception of the chemistry of airs. Watt's long-held belief about the conversion of water into air if high enough temperatures could be attained had meteorological significance. Recall that when he advised Boulton of this idea in December 1782 he noted that it solved the problem of the source of atmospheric air, as Priestley had already observed.

The work that Priestley did for Boulton and Watt to explore the possibilities of using other airs for engines was also conceptually linked to this natural economy. Boulton and Watt were alert to the possibilities of using the capability of fluids, or even solids, to release and reabsorb air to produce a vacuum and derive mechanical force thereby. In the letter in which he explained to Watt that he was setting Priestley to work on the question, Boulton gave details of what had impelled him to do so. One circumstance had been seeing in London 'a Glass Vessell filld w[i]th >⊖ air which became as instantaneously condensed when brought into contact with ▽ as steam w[oul]d have been, & the same thing happens with ⨸^ air…'[43] Boulton explained further, with accompanying diagram (see Figure 5.1)

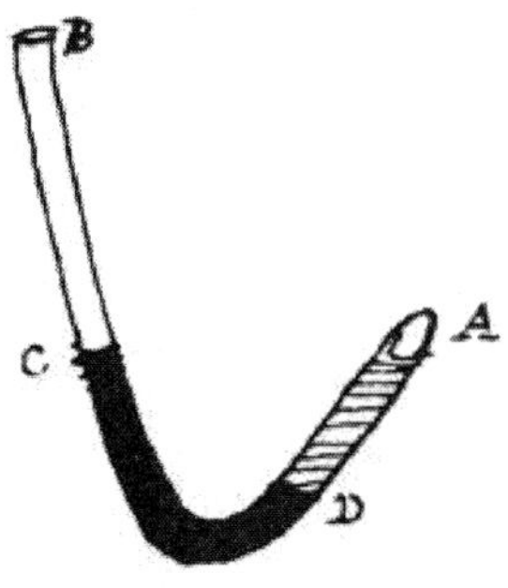

Now if you take a Glass tube hermetically seald at A & open at B – let the space between C & D be filld with mercury & the space between D&A be filled w[i]th water into which project as much ⨸^ air or >⊖ air as will saturate that ▽ … The moment you apply heat to the water, it gives out the air & ye instant you cool it that instant it is absorbed again by ye ▽. Heat it again & it gives out air so as to raise the mercury to a considerable height…[44]

Figure 5.1. Diagram from Boulton to Watt, 21 July 1781. (Archives of Soho, MS3147/3/5, by Kind Permission Birmingham City Archives)

Besides seeing the London experiment, Boulton had also been stimulated by a conversation with 'Simcox'[45] who 'mentioned to me the Idea of detaching fixable air from chalk & then in an instant reabsorbing it'. Boulton reported being alarmed by this prospect until he was convinced of the 'impracticability' of it. He remained convinced, however, that what could not be accomplished with solids might be with fluids, and argued that they should take out a patent against the eventuality:

> a patent for making Engines by means of alternately emiting & reabsorbing all & severaly the different airs, Gasses, permanently elastic fluids, or by what soever other name they might be calld, seting forth that whereas there are several substances that do contain or are capable of containing several sorts of air acids &c in a fixed or solid state & that such airs &c are readily detached from, or are emitted by such substances upon being heated & again upon being coold are readily reabsorbed by the s[ai]d substance & thereby a vacuum may be obtained & Mechanical force derived.[46]

In the case of the emission and reabsorption of airs by solid substances the links to geological processes are clear. While Watt's reply calmed Boulton down, convincing him that such schemes, whether using solids or fluids, were impracticable, the fact that they could be contemplated at all confirms the existence in Watt's circle of the connections I am claiming between thinking about engines, about the chemistry of airs, geology and meteorology.

Another example, this time of the perceived connection between steam experiments and geology, comes twenty years later, when, in retirement, Watt repeated many of his steam experiments, much of the actual work being done by his assistant, John Southern. The papers dealing with Southern's calculations of the elasticity of steam dating from 1803 contain an interesting document annotated 'Considerations upon the power of steam to produce Earthquakes by Mr Southern 1803'.[47] This is worth considering in a little detail.

The 'Considerations' begins with an attempt by Southern (one of many in these papers) to establish the series that elasticities of steam follow. It is recorded that 'J. S. found that the specific gravity of steam of atmospheric elasticity was 2000 ... times lighter than water. [S]upposing then that if its elasticity is increased its requisite density increases in the same ratio, it would be when equal to 2000 atmospheres as dense as water'. The question is then asked, what temperature would give this? The calculation results in an answer of '1107 of Farenheit [*sic*]'. It then becomes apparent that this work is preparatory to dealing with a question about the Earth – whether 'the powers of steam' can be 'contemplated with the view to their sufficientcy [sic] to produce Earthquakes, Volcanoes &c'. The answer provided is that saturated steam could never achieve the elasticities to do this, but that unsaturated steam might: 'though one may not look for saturated steam denser than water perhaps & consequently only of the elasticity just mentioned[,] unsaturated steam may be adequate to these effects'. The chain of reasoning arrives at the proposition that a cubic mile of water raised to a temperature of 5100° within a cavity in the Earth of 32 cubic miles would 'fill the ... cavity with steam that would press on the whole of the surface with a force of 400 miles perpendicular of water. This might I think produce earthquakes and volcanoes'.

Neither the detailed merits of this analysis, nor where it might fit in the geological controversies of the time, need detain us here.[48] The key point is Southern's and Watt's belief that the characteristics of steam (that is the chemistry of water and heat) were relevant to the solution of the problems of both the steam engine and the Earth-machine. Knowing this, the article 'Steam' by John Robison in the third edition of the *Encyclopaedia Britannica* begins to look less eccentric than it otherwise might. When Robison segued from a sober discussion of the boiling of water in a kettle into rhapsodies on earthquakes and other geological phenomena[49] he was in fact reflecting a natural philosophical worldview still common among his circle and likely shared by Watt himself.

Steam in *The Botanic Garden* and 'Sentimental Empiricism'

Another member of the Lunar Society, Erasmus Darwin, took an abiding interest in the steam engine, an interest at once visionary, poetic and scientific. We find in sources produced by and linked to him, further confirmation of the chemical nature of the invention as seen by Boulton and Watt's contemporaries and also of its close association with meteorological ideas and understandings.

Even before Watt came on the scene Darwin and Boulton were discussing steam engines. As the story goes, Boulton's worries about the possible lack of water power to sustain his manufactory led him to serious investigation of steam engines from the mid-1760s. He pursued his own steam project until he met Watt in 1768. Writing to Boulton in December 1765, Darwin expressed curiosity about how Boulton's steam engine had been received by Dr Franklin and what Franklin's observations on it had been. Darwin, Boulton and Dr William Small had been discussing boilers and their optimal design. This question involved them in considering the phenomenon of evaporation. Darwin wanted to know 'your final opinion and Dr Small's on the important Question, whether Evaporation is as the Surface of boiling Water, or not? – or if it be as the Surface of the Vessel, exposed to the Fire, which I rather suspect'. He continued:

> For if you boil Water in a Florence-Flask, you see bubbles arising from the Bottom, these are Steam, as they mount they are condensed again by the cold supernatant Fluid, and their Sides clapping together make that Noise called Simmerring. 2dly When Water boils in the Æolipile or closed Tea-Kettle, what a Quantity of Steam? – is this only as the upper Surface? I humbly conceive not – but as that Surface in contact with the Fire.[50]

This question of *where* the boiling took place was crucial to the design of the boiler of the steam engine. Darwin sought from Boulton his, and Small's, view on 'the best way of constructing the Boiler for the above Purposes'.[51] The question of which variable related to the quantity of steam produced was also of theoretical significance. Darwin was interested in what Franklin thought because Franklin subscribed to the solution theory of evaporation. Solution theory would be compatible with the proportionality of quantity of steam to surface area of the water to be boiled; solution, in any case, requiring solvent (air) and solute (water) to be in contact. The quantity of steam produced being proportional to the surface area of the boiler in contact with the fire would be consistent with the idea that steam is compounded from water and heat, or the matter of fire, since this could occur, and did on that theory occur, without contact with the air. At about the same time, but apparently independently, Watt was also considering 'boyler' design in much the same light.[52]

Over twenty years later Darwin's interest in evaporation and 'devaporation' persisted. In his celebrated paper on 'frigorific experiments' published in the *Philosophical Transactions*[53] he drew close analogies between what goes on in the steam engine and in meteorological processes. Darwin's central concern in this paper was to show that many meteorological phenomena could be accounted for by applying a basic insight to the issue of movements of air in the atmosphere. That basic insight was that elastic fluids, when mechanically expanded, withdraw heat from their surroundings. In considering the application of this to the 'devaporation of aerial Moisture' Darwin suggested that when the barometer drops, for whatever reason, the lower parts of the air become expanded and, in expanding, extract heat from the vapour contained in them. The vapour condenses and it is precipitated in showers. But a remarkable and little-noticed effect is that 'the deduction of a small quantity of heat from a cloud or province of vapour, compared with the quantity of heat which was necessary to raise the vapour from water, will devaporate the whole'.[54] Darwin explained this as follows: if 'a small part of a province of vapour' is suddenly condensed, a vacuum is formed and the proximate bodies of vapour expand into it. Thus 'a large area of vapour, perhaps of many miles in circumference, becomes more or less expanded', by this expansion cold is produced, 'and the whole is devaporated'. Then comes the analogy with the steam engine:

> This very circumstance exactly takes place in the famous steam-engine of Mess. WATT and BOULTON; which, from the happy combination of chemical and mechanical power, may justly be esteemed the first machine of human invention. In this excellent machine, after the cylinder is filled with steam, a communication is opened between this reservoir of steam and a small cell, which is kept cold by surrounding water, and free from air ... What then happens? The corner of the steam in the cylinder next to this vacuum (with which it now communicates) rushes into it, and the whole steam in the cylinder is thus suddenly expanded and instantly devaporated...[55]

Thus, according to Darwin, the process that enables Watt's separate condenser to work is the same as that which operates, for example, at the beginning of thunderstorms, when 'a small black cloud at first appears, in a few minutes the whole heaven is covered with condensing vapour'. It is worth noting also that in the above quotation Darwin confirms the perception of the improved steam engine as a chemical device. By referring to it as the 'first machine of human invention' he means, presumably, the first in rank rather than the first in time. This pre-eminence is awarded because of the machine's 'happy combination of chemical and mechanical power', the latter deriving from, and being used in concert with, the power derived from application of the knowledge of the chemistry of steam.[56]

Darwin had taken a long and close interest in Watt's steam-engine improvements and famously gave the steam engine a place in his poem *The Botanic Garden*:

> NYMPHS! You erewhile on simmering cauldrons play'd,
> And call'd delighted SAVERY to your aid;
> Bade round the youth explosive STEAM aspire
> In gathering clouds, and wing'd the wave with fire;
> Bade with cold streams the quick expansion stop.
> And sunk the immense of vapour to a drop. –
> Press'd by the ponderous air the Piston falls
> Resistless, sliding through its iron walls;
> Quick moves the balanced beam, of giant-birth,
> Wields his large limbs, and nodding shakes the earth.[57]

The first thing to note about this passage is the tenor of the language, at once meteorological, mechanical and animist. The immediate notes on the passage recount the history of the steam engine from Savery through Worcester and Newcomen to Watt and Boulton. They also refer to its applications and future possibilities. As we saw in Chapter 2, material for Additional Note XI on the steam engine was contributed by Boulton and Watt themselves.[58] The Note as published contains a version of the 'Perfect Engine' story intermediate between Watt's letters of the time describing the invention and the 'Plain Story' that originated in the 1796 patent trials and was one of the two key versions to be propagated in the nineteenth century.

More generally in *The Botanic Garden* we can discern Darwin's partial assimilation of Lavoisian chemistry in developing his economy of nature. Darwin accepts the Lavoisian account of the decomposition and composition of water. He puts the explosive combination of inflammable air and oxygene (which he still also denominates as dephlogisticated air) to work in various parts of the natural economy, including the upper reaches of the atmosphere, where their sparking he considered likely to produce instant summer rain, and in the formation of fissures in the Earth during its early history. However, Darwin also relied heavily on the 'doctrine of elementary heat', which he identified with Joseph Black and his followers. Darwin stated:

> A certain quantity of heat seems to be combined with all bodies besides the sensible quantity which gravitates like the electric fluid amongst them. This combined heat or latent heat of Dr. Black, when set at liberty by fermentation, inflammation, crystallization, freezing, or other chemical attractions producing new combinations, passes as a fluid element into the surrounding bodies. And by thawing, diffusion of neutral salts

> in water, melting, and other chemical solutions, a portion of heat is attracted from the bodies in vicinity and enters into or becomes combined with the new solutions.[59]

He proffered as an additional reason for belief in the doctrine his ideas about the absorption or release of heat when air is mechanically expanded or condensed. Darwin, for all he was the most accepting among the Lunar group of Lavoisier's new chemistry still conceived it within what Riskin[60] has called a 'sentimental empiricism', a view of the economy of nature which he still shared, I suggest, with James Hutton and with Watt and his other Lunar Society colleagues, and which was quite alien to the Lavoisian scheme. Although Darwin writes of 'calorique' as matter of heat in a way that assumes it is interchangeable with the doctrine of elementary heat in the Scottish tradition, we must regard this as obfuscating very real differences between them.[61] In fact, through his poetry Darwin gave us the most evocative clue to this common quality in his own cosmology and that of Watt, Hutton, De Luc and other members of their circle; this approach prevailed over their substantive differences on specific issues in natural philosophy. It treated the steam engine as a local manifestation of a kind of chemical power that was ubiquitous in the economy of nature.

Watt's Steam Experiments

Now let us look finally, and directly, at Watt's experiments on steam and how they related to his steam-engine work. A small notebook in the Watt archives, most of it in Watt's hand, contains notes about steam experiments, conducted in

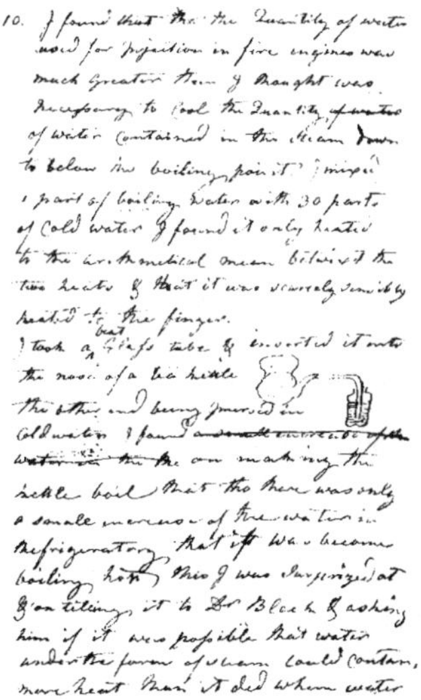
10. I found that the Quantity of water
used for injection in fire engines was
much Greater than I thought was
necessary to cool the Quantity
of water contained in the steam down
to below the boiling point. I mixed
1 part of boiling water with 30 parts
of Cold water I found it only heated
to the arithmetical mean betwixt the
two heats & that it was scarcely sensibly
heated to the finger.
I took a bent Glass tube & inverted it into
the nose of a tea kettle
the other end being immersed in
Cold water I found on making the
kettle boil that tho there was only
a small increase of the water in
the refrigeratory that it was become
boiling hot. This I was surprized at
& on telling it to Dr Black & asking
him if it was possible that water
under the form of steam could contain
more heat than it did when water

Figure 5.2. Page from Watt's 1765 Experiments, with Kettle Sketch. [Reproduced by kind permission of Birmingham City Archives]

1765 or earlier and in the early 1780s.[62] Notes were added in 1814 when Watt returned to these questions. These are the real kettle experiments, which were arguably the inspiration for the myth of the kettle discussed in Chapter 1. On a page headed '1765' there begins a long account of Watt's first experiments on steam. Watt recalls finding that a greater quantity of injection water than expected was required to condense steam in the cylinder of a steam engine. In particular, more water was needed to condense the steam than should have been required to cool the quantity of water in the steam down below boiling point. Watt recalled (see Figure 5.2):

> I took a bent Glass tube & inverted it into the nose of a tea kettle, the other end being immersed in Cold water. I found on making the kettle boil that tho' there was only a small increase of the water in the frigeratory that it was become boiling hot. This I was surprised at & on telling it to Dr Black & asking him if it was possible that water under the form of steam could contain more heat than it did when water He told me that had long been a tenet of his...[63]

In this experiment, Watt was arriving independently, sometime before 1765, at the fact of the latent heat of condensation of steam.

What was the relationship between these experiments and Watt's improvements in the steam engine? We need to recall what the key improvement of the separate condenser consisted of. In the Newcomen Engine cold water was injected into the cylinder on every stroke, condensing the steam and creating a partial vacuum so that the atmosphere forced the piston down. More steam was injected, and so the cycle continued.[64] The idea of latent heat explained the initially surprising amount of cold water needed to condense the steam in the cylinder. However, latent heat was not central to the engine inefficiency to which Watt addressed himself. The problem was that during condensation the cylinder itself was cooled by the injection water. The injected steam in the next cycle had to reheat the cylinder. The repeated reheating of the cylinder consumed a lot of fuel. When Watt recognized this inefficiency, his first idea was to construct the cylinder of wood, or some other material with a lower specific heat, and he conducted trials with various materials. Subsequently, Watt came up with the idea of the separate condenser. In this arrangement the cylinder was not repeatedly cooled and reheated. The fuel saving so achieved was enormous and was the technical basis of the firm of Boulton & Watt's success.

The relationship between Watt's investigations of the properties of steam and his improvements to the steam engine is not straightforward. Watt investigated a number of properties of steam. Following Black, he also conducted experiments on the boiling of water at lower pressures. This was linked to the creation of 'back pressure' in the cylinder when the injection water boiled as a result of the partial vacuum in the cylinder and the increased temperature produced by release of the latent heat of condensation of the steam. Arthur Donovan points out that Watt's study of this 'back pressure' led to the formulation of the problem to which the separate condenser would be a solution. This was expressed in terms of the departure of real from perfect engines. As Watt put it:

> to make a perfect steam-engine, it was necessary that the cylinder should be always as hot as the steam which entered it, and that the steam should be cooled down below 100° in order to exert its full powers. The gain by such construction would be double: – first, no steam would be condensed on entering the cylinder; and secondly the power exerted would be greater as the steam was more cooled.[65]

It was at this point that the idea of the separate condenser occurred to Watt. This idea depended on yet another property of steam, namely its 'elasticity' as a vapour. Watt realized that this property of elasticity meant that if an evacuated vessel were linked to the main cylinder, steam would rush into it. Once captured, the steam could be condensed by injection water without having to cool the cylinder significantly.

Here is an appropriate place to discuss Richard Hills's important and influential reconstruction of Watt's route to the separate condenser.[66] Hills presents the concept of the 'perfect engine' as the guiding idea here. His account runs the risk, in my view, of assimilating the concept to proto-thermodynamic physical ideas. I want to argue that for Watt in 1769 preventing the waste of steam, reducing it theoretically to zero, was conceptualized chemically, but first let me set out briefly what I take to be Hills's argument.

Hills begins with the famous story that during the winter of 1763–4 Watt, who was then an instrument maker with a shop at Glasgow University, was asked by Professor Anderson to repair a model of a Newcomen steam engine belonging to the natural philosophy class. According to Watt's account he was surprised that the boiler on this model could not supply enough steam to keep the engine going. Watt set about studying what had been written about steam engines to try to understand the basic principles of their operation. Hills tells us that by September 1765 Watt used the term 'perfect engine' in a letter to Dr Lind:

> I have tried my small model of my perfect engine, which hitherto answers expectation, and gives great, I may say greatest, hopes of success (for certainty could not be called hope), in greater model now far advanced; in short, I expect almost totally to prevent waste of steam, and consequently to bring the machine to its *ultimatum*.

The standard of perfection in an engine was thus conceptualized by Watt in terms of the degree to which steam was wasted. In a perfect engine, none was. The concept gave Watt a standard against which to judge any working engine.

So Watt was on the hunt for sources of waste. He knew that in the Newcomen engine the cylinder had to be re-heated at each stroke, that is, a large quantity of steam had to enter, and give up its heat to, the cylinder before the latter became hot enough to be filled with steam that remained as steam. The injection of cold water into the cylinder in order to condense the steam within it and so create a vacuum was responsible for the cooling of the cylinder. Watt's experiments convinced him that this was the single largest source of wastage of steam. Hills' account beyond this traces Watt's inspiration of the separate condenser, and how that met the basic criteria that a 'perfect engine' demanded. He also discusses other problems with the Newcomen engine and how Watt dealt with those. What are we to make of this account?

Watt certainly used the term 'perfect engine' at the time of his work on the improvement of the steam engine. He also deployed it in retrospective accounts, as we saw in Chapter 2, and it became popular among the first generation of historians, including James Patrick Muirhead. But what work does the concept do? Once it establishes the idea of a perfect engine (which is itself little more than the idea that waste is occurring and that in the best sort of engine it would not) the answer has to be 'not much'. Hills's own account of Watt's work in tackling various sources and aspects of waste does not need to invoke the idea at all, and it does not do so. That account is, in fact, all about heat, that is to say, about chemistry in the then contemporary understanding, certainly in Watt's understanding. At one stage, in his most recent treatment of the subject, Hills acknowledges this very point:

> [Watt] ... could never know that his concept of the perfect engine was flawed. He considered that heat was some sort of chemical substance and regarded the engine as working by the pressure of steam. He thought it was essential to keep the cylinder as hot as the incoming steam all the time to avoid condensation and hence loss of pressure. But he could not know that heat is a form of energy which, when it is transformed into work and drives machinery, must decrease in temperature with the result that the cylinder must become colder during the working stroke. The reason for this dilemma was recognised only in the middle of the nineteenth century with the rise of thermodynamics...[67]

Whilst ostensibly escaping presentism here, Hills's account actually embraces it in the judgement that Watt's understanding of the 'perfect engine' concept was 'flawed' rather than merely different. Watt's concept was only 'flawed' on anachronistic criteria. It was precisely the chemical interpretation that distinguished Watt's 'perfect engine' concept from that of nineteenth-century thermodynamics. Thus even if we grant the centrality of the idea of the 'perfect engine' to Watt's reasoning, in order to avoid anachronism we still have to conceive of that perfection in terms of the chemistry of heat. When heat was thought of as a substance involved in the chemical formation of steam, as Watt thought of it, then it was only necessary to supply enough heat in order to achieve perfection. I think that the concept of the 'perfect engine' has been seized upon for its apparent summative potential. The term evokes ideas clearly or dimly recollected of the 'ideal engine' from physics and thermodynamics. To use the term 'perfect engine' as the organizing principle for an account of Watt's steam engine is to conjure up other ideas such as the bulk properties of steam. Logically, given what we *now* believe about latent heat, namely that it is a non-chemical phenomenon, the conception of the perfect engine was a major inspiration for the separate condenser idea. Watt's reasoning did not involve the chemistry of heat directly, and it is tempting to see it as a kind of proto-thermodynamics, given that it *seems* to deal with the bulk thermal properties of a 'working substance'. But the essence of a working substance in that sense is that it operates as a vehicle of heat transfer and nothing more. The theory of the heat engine becomes independent of the working substance. This could never be the case for Watt since his conception of the steam engine depended so intimately on the chemical composition of the steam, which varied according to the amount of heat chemically combined with it, that is latent within it.

Ben Marsden has the right idea when he warns his readers against importing nineteenth-century concepts into Watt's time. Certainly Watt was not trying to maximize 'energy' output or conversion, he was seeking to minimize waste.[68] But the concept of 'perfect engine' is susceptible of similar misunderstandings. Once the meaning of the term has been explained as an engine that does not waste steam it carries no further meaning. As an abstraction, unlike 'ideal engine', it does no further work since it has no further content.

The story of Watt's perfect engine can stand. As we have seen, Watt did use the term at a crucial juncture to describe to others what he was doing. The story is undoubtedly true when that term is understood as Watt and his contemporaries might have understood it. It may well be, as Hills suggests, that the term was borrowed from other contemporary areas of technology practice. Perhaps the concept guided Watt to the idea that the central problem was that of the waste of steam. All these suggestions are fine and useful. The problem arises when Watt's 'perfect engine' elides historiographically into notions of 'ideal engine' of the sort that was to become important in the nineteenth century. When that happens we think we have understood Watt, but we have not. There is still work to do.

Let us return now to Watt's experiments on steam and latent heat. In late February and early March 1781, Watt resumed experiments on latent heat:

> The straight end of a Pipe was fixed on the Spout of a Tea kettle from which the pipe inclined upwards so that the Bent end was about 2 feet higher than the Spout of the Tea kettle. The Tea Kettle was filled with Water halfway up the entry of the spout. The Lid was fixed Steam tight with some oat meal dough.[69]

The pipe from the tea kettle led into a pan containing water. Watt measured the starting heat, and weight, of the water in the pan. Then steam from the kettle was

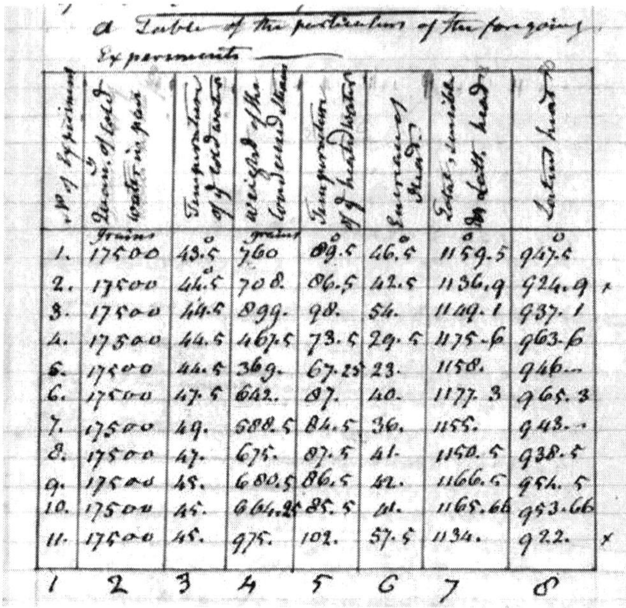

A Table of the particulars of the foregoing Experiments

No of Experiment	Quan. of cold water in grains	Temperature of cold water	weight of the condensed Steam	Temperature of heated water	Increase of heat	Latent, sensible or both heat	Latent heat	
	grains	°	grains	°	°	°	°	
1.	17500	43.5	760	89.5	46.5	1159.5	947.5	
2.	17500	44.5	708.	86.5	42.5	1136.9	924.9	x
3.	17500	44.5	899.	98.	54.	1149.1	937.1	
4.	17500	44.5	467.5	73.5	29.5	1175.6	963.6	
5.	17500	44.5	369.	67.25	23.	1158.	946...	
6.	17500	47.5	642.	87.	40.	1177.3	965.3	
7.	17500	49.	588.5	84.5	36.	1155.	943...	
8.	17500	47.	675.	87.5	41.	1150.5	938.5	
9.	17500	45.	680.5	86.5	42.	1166.5	954.5	
10.	17500	45.	664.25	85.5	41.	1165.66	953.66	
11.	17500	45.	975.	102.	57.5	1134.	922.	x
1	2	3	4	5	6	7	8	

Figure 5.3. Watt's Tabulation of Experiments on Heat
[Reproduced by kind permission of Birmingham City Archives]

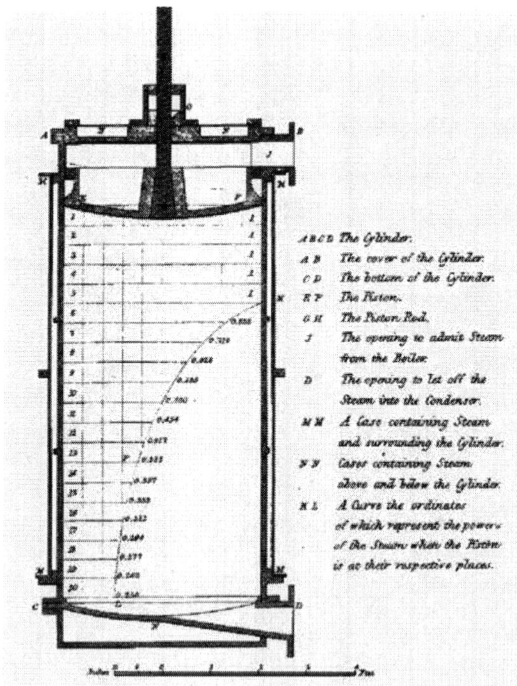

Figure 5.4. Watt's Expansive Working, from 1782 Patent. [From: J.P. Muirhead, *The Origin and Progress of the Mechanical Inventions of James Watt* (London, 1854), vol. 3, plate viii].

allowed to condense in the pan and the finishing heat and weight of the water in the pan were recorded. The experiment was repeated a number of times. When he reviewed these experiments, in 1814, Watt tabulated them[70] as in Figure 5.3.

It can be seen that Watt calculated the 'Total, sensible & Latt. Heat' (Figure 5.3, column 7). In the 1780s, Watt believed that the sum of the latent and sensible heats was a constant. This idea is the basis of what was known into the nineteenth century as 'Watt's Law'. It was as we have seen, important to Watt's thinking about the composition of water in the early 1780s and was also closely related to his exploration of the principle of 'expansive working' of the steam engine, which he explained in a patent taken out in 1782. Expansive working was ultimately about saving steam. It was about the 'steam economy' in the steam engine. Watt observed that he could refrain from injecting steam into the cylinder throughout the cycle. Instead he considered what would happen if the injection of steam ceased one quarter of the way through the expansion stroke. The steam would continue to expand at decreasing pressure. Watt worked out what the average pressure would be using such expansive working, and he came up with a figure of 0.579 (where 1 is the pressure when steam is being injected). The average pressure using expansive working was more or less halved. But Watt found that the amount of steam used was a quarter of the amount used in normal working.

Note that in the diagram representing this taken from the 1782 patent (Figure 5.4) Watt describes the line KL as representing 'the powers of the Steam when the Piston is at their respective places'. He is tracing the declining elasticity of the steam and to him this meant the rate at which the combined latent heat in steam was being lost and converted into sensible heat. This is where 'Watt's Law' comes in. If it holds, then, given a measure of the sensible heat, the amount of latent heat, and therefore the elasticity of the steam can be determined.

Conclusion

So, how should we summarize the chemistry of the steam engine? We see that the link between latent heat and Watt's improvements was a complex and *contingent* one. Donald Fleming went too far when he rejected the link entirely because it was not direct.[71] Certainly the innovation of the separate condenser relied for most of its gains in efficiency on the fact that the cylinder was not being repeatedly heated and cooled. Those gains would be achieved regardless of the existence of latent heat. Nevertheless, the formulation of the problem of back pressure to which the separate condenser was a solution, did owe a lot to Watt's understanding of latent heat and to other of his experiments on the properties of steam. Those properties were marginal to Watt's key initial improvement, but not wholly irrelevant to it. They were not the proximate cause of the key improvement but were certainly part of his overall understanding of what was going on in the engine and a clue to the problem of the waste of steam and high fuel consumption.[72] The later development of the idea of expansive working, however, and the pursuit of 'steam economy', did involve Watt in thinking much more closely about what was happening in the cylinder. Here the composition of the steam in terms of how much latent heat it contained in various circumstances became crucial.

What I wish to emphasize is that throughout all these experiments Watt understood the activity of steam as being a result of its chemical character as a *compound* of water and heat. The elasticity of steam was the basis of the force that it exerted in order to drive the piston and was also the property that allowed the separate condenser to work. The degree of elasticity of the steam in the cylinder of the engine depended directly on the amount of latent heat combined in the steam. In short, all key aspects of the engine's working were the product of steam's *chemical* character. Watt saw the improvement of the engine as a matter of learning more about the chemical changes occurring in the steam so as to be able to take manipulative advantage of them. The narrative giving pride of place to Watt's concept of the 'perfect engine' is, when properly understood, compatible with this position. But I think it gives hostage to historiographical fortune by encouraging anachronistic analogies with ideal engines and diverting our attention from the chemical understanding of the steam engine that underlay Watt's innovations.

6 THE INDICATOR UNDERSTOOD, OR WHY WATT WAS NOT A PROTO-THERMODYNAMICIST

Introduction

Later nineteenth-century histories of the steam engine were most often written by practising engineers and scientists. They exhibited the common tendency of practitioners to write 'Whig' history, that is, to interpret past historical actors and actions in modern terms. This genre of literature, then, assimilated Watt to the modern traditions and conceptions of thermodynamics. In its turn this later nineteenth-century literature has left its mark on modern historical writings.

I am concerned in this chapter with two processes of assimilation of Watt to thermodynamics. The first involves conceptual assimilation of Watt to mid-nineteenth-century understandings of 'energy'. A typical example of conceptual assimilation is an essay by Keith J. Laidler, which presents Watt and his engine as practical progenitors of that science. Certainly the steam engine was an important resource for those developing thermodynamic understandings. But to picture Watt and his engine 'pushing' in that direction is problematic.[1] Even sophisticated and nuanced treatments of the issue fall into a similar trap. The eminent historian of technology, Donald Cardwell, implicitly placed the work of Black and Watt on heat into a tradition of the 'physics' of heat, a line leading to the 'Rise of Thermodynamics'.[2] I have already argued that the excellent and detailed account of the origins of Watt's key invention by Richard L. Hills[3] is also on dangerous territory when deploying the concept of the 'perfect engine' in telling that history. Although the term does come from Watt's own accounts of how he arrived at his invention, and therefore must in some sense be reliable,[4] what the term meant to Watt is all too easily lost sight of and elided into later visualizations of ideal heat engines that every modern student learns about as the basis of thermodynamics. The second process of assimilation involved the steam indicator, a device that, because of its mid-nineteenth-century involvement with thermodynamic theory, was often taken as some kind of natural bridge between that theory and Watt, who invented the instrument. As I will show, the rather mysterious, chequered and just plain misunderstood, history of this device encouraged a 'telescoping' of its later and early history. That telescoping in its turn facilitated an easy elision between Watt's conceptual world and that of the thermodynamicists.

Watt, Heat and Energy

The elisions which concern me here, and which I will reveal in the case of understandings of the indicator, have a long history. As we saw in Chapter 2, some of them were perpetrated during Watt's self-fashioning in the early nineteenth century and entered the historiography of the Watt camp including the major works by J. P. Muirhead. From there, as well as from their own technical literature, the thermodynamicists used them to understand and shape their own history. A particularly concentrated and revealing source is the sequence of Watt Anniversary Addresses given at the Greenock Philosophical Society from 1865. There was a long tradition in Greenock of celebrating the anniversary of Watt's birth dating back to the dinners and lunches of the Watt Club from the 1820s. The formation of the Greenock Philosophical Society in 1861, which absorbed the membership of the Watt Club, placed these affairs on a more formal basis and involved addresses given by numerous prominent engineers and scientists.[5] The occasions, like the Watt Anniversary Dinners long held in Glasgow itself under the auspices of the Institution of Engineers and Shipbuilders of Scotland, were important to, and supported by, the west of Scotland engineering and shipbuilding community whose academic centre was the University of Glasgow. Given the stature and importance of that community to the world of engineering in the later nineteenth century, the Watt Anniversary became much more than a local celebration – a fact attested to by the speakers that it attracted.[6]

The first Watt Anniversary Address was in fact given by none other than James Joule in 1865. Joule's treatment of Watt's ideas on heat is worth recounting at some length both for our immediate purpose of seeing how the elisions I identify were made by him, and for more general interest, since to my knowledge its content has never been discussed before. Donald Cardwell, in his biography of Joule, noticed the Address and observed that it was a truly rare occasion, being probably the only public lecture that Joule ever gave to an audience other than a small local gathering. Cardwell does not, however, venture into the contents of the Address.[7]

Joule spoke to the Greenock Philosophical Society in the Watt Institution on 19 January 1865, 'On Some Facts in the Science of Heat Developed Since the Time of Watt'.[8] The bulk of the lecture was given over to a description of Joule's own, already famous, experimental demonstration of the mechanical equivalent of heat and to the work of his friends William Thomson and Macquorn Rankine in developing the science of thermodynamics. But Joule could hardly avoid a historical sketch of developments concerning heat between Watt's time and the mid-nineteenth century. The way that he positioned Watt, or failed to position him, in that sketch is revealing. Joule noted the obvious first – that they were assembled to mark the anniversary of the birth of 'the greatest mechanician that this or any other country has ever produced'. The term 'mechanician', especially from the lips of a man famous for determining the mechanical equivalent of heat, seems to place Watt preferentially as a student and exponent of the mechanical side, rather than the heat side, of that equivalence. Joule also observed that the purpose of the lecture was to 'reflect on some of those facts of physical science which are intimately connected with the discoveries of Watt, and which give a rational explanation of the operation of that wonderful machine which will for all time be associated with his name'. Once again, Watt is associated with the engine, but, at this stage at least, only equivocally with the 'rational explanation' of its operation.

Joule then launched into an account of the history of ideas about the nature of heat. He ventured that the conception of heat as elemental, held by the Ancients, was vastly improved in the hands of Bacon, Newton and other great men of the seventeenth century who considered heat as 'a motion among the particles of matter'. This, together with Newton's announcement, as Joule depicted it, of the principle of the conservation of energy, laid the foundation of the science of heat. Then, in the eighteenth century, came a retrograde step. Here is Joule's account of the nature of that step and the reasons for it:

> The overthrow of Alchemy and the rise of Chemistry, with its definite elements and combining proportions, seem, singularly enough, to have caused a retrograde movement in the sister science. Still, as more precise ideas began to prevail respecting the individuality of different species of matter, it was natural enough to attempt to establish a material theory of the then called 'imponderables', more especially as such a theory was found to give a very intelligible explanation of many calorific phenomena; such, for instance, as the absorption of heat when a body assumes another form, as, for example, when ice becomes water, or when water becomes steam. The thermal laws in connexion with such changes of state were demonstrated by the celebrated Black, and nothing could appear more natural than to suppose with him that the change of condition was owing to the union of the substance with Caloric, which latter, being neutralized by the combination, ceased to act on the senses, and was therefore called 'latent heat'.[9]

This is the closest that we come in the entire lecture to a treatment of Watt's ideas about heat. Although Watt's name is not mentioned at this point, his presence in the thought is unmistakable. He was the person for whom 'nothing could ever appear more natural' than to adopt, following Black, the view that changes of state involved the chemical union of substances with caloric. Watt is hereby *excused* for his views on the nature and role of heat. He was swept up by a whole tradition that had gone astray – the science of imponderables. He followed Black. In any case both Black and Watt transcended this situation. They had the 'thermal laws', which were disengageable, Joule implied, from the question of the nature of heat. Such a view of the 'thermal laws' was in direct line of descent from the general strategy of Watt's self-fashioning in the later years of his life.

The rest of his history of heat involves Joule paying obeisance to the work of Rumford and Davy, which he regarded as decisive in favour of the mechanical theory of heat. However, 'frivolous objections' led to it being forgotten so that 'the several naturalists who more recently have recommenced the enquiry have done so quite independently, and on ground of their own selection'. Watt disappears from the lecture for a while, as Joule recounts his own work and that of Thomson and Rankine. But, in concluding, Joule says this:

> I must now close this very imperfect outline of a few of the discoveries which have taken place since the time of that great man whose birth we are now commemorating. I thank you for the opportunity you have given me of discoursing on results which would have been most interesting to Watt, *and for which he himself largely prepared the way*. And I trust you will permit me also to express my satisfaction at having had it in my power to speak in Scotland of a theory which Scotland was the earliest country to welcome, and which your countrymen have done more than any others to complete and extend.[10]

So, even as Watt's own ideas about heat and the steam engine are quietly acknowledged as deflected from the main line of development, he is also assimilated to that main line. This is done by picturing Watt as preparing the way for it (presumably through the application of the 'thermal laws' to the design of the steam engine from which the likes of Rankine has, as it were, reverse engineered the laws of thermodynamics), as being (virtually!) interested in it, and as being part of that company of Scotsmen who had advanced it!

Henry Dyer's Watt Lecture of 1889 on 'The Steam Engine since the Days of Watt' confronted Watt's ideas about heat more directly. He did this through a consideration of Watt's invention of the separate condenser. Dyer, following and quoting from Watt's own corrective account in Robison's *A System of Mechanical Philosophy* explained that while it would have been possible for Watt to derive his engine improvements from the concept of latent heat, this 'does not seem to have been the way in which this discovery was developed'.[11] In a prior exposition of Black's doctrine of latent heat, Dyer stated that 'Black first showed ... that heat may enter a body and cause it to change its molecular condition, while the temperature as indicated by the thermometer remains constant, in other words that the heat becomes *latent*'. The phrase 'molecular condition' is, of course, a modernization of Black's actual conception. In fact Dyer goes on to give this rendering of what Black and Watt believed. According to the material theory of heat, he says, it was supposed:

> that heat was an imponderable, gaseous fluid, which, entered the pores of bodies, and caused them to expand, and which combined with their substance and became latent or hidden when they pass from the solid into the liquid state, or from the liquid to the vaporous, and which re-appears when the passage is reversed. Although this theory is no longer held, still the name *latent* is retained, as it expresses clearly enough what actually takes place.[12]

Whether through ignorance or design, Dyer here is attributing to Black and Watt a curious combination of a *chemical* material theory of heat and an Irvinist, non-chemical material theory of the sort that became popular in the early nineteenth century. This is a good example of the sorts of elisions that converted Watt's and Black's theories of heat into something assimilable to nineteenth-century physical conceptions. At the end of the historical excursion Dyer makes an interesting and revealing statement that his main reason for taking the reader on it was not to weigh the relative merits of Black and Watt but rather to illustrate the fact that when we study the history of the steam engine we find that

> the great improvements which have been made in it have not always been the result of an exact appreciation of the full extent of the laws which govern them, but rather of clear-sighted observation of experiments on the part of men gifted with mechanical genius, who have perceived the direction in which these laws tend...[13]

Dyer's solution to the dilemma about what to do with the disjunction between Watt's understanding of heat in relation to the steam engine and that given by nineteenth-century physics was to elevate that disjunction to the level of a historical generalization about the relationship between theory and practice. Watt's accomplishment was the product of his application of clear-sighted experiment and mechanical genius assisted by his perception of the 'tendency' of natural law. We are not far here from Cardwell's depiction of Watt as a proto-thermodynamicist, as we see when Dyer precedes his discussion of Carnot's work that 'laid the foundations of the thermodynamical theory' of heat engines with the observation that 'Watt seems to have perceived, if not the law, at least some of the conditions of maximum efficiency of steam engines considered as heat engines'. Dyer simply assumed that Watt had the modern concept of a 'heat engine'.

Similar anachronistic treatments of Watt pervaded more accessible accounts. Thus J. D. Forbes in his Preliminary Dissertation to the eighth edition of the *Encyclopaedia Britannica*, implied Watt's knowledge of 'energy'. Discussing the Cornish engines of the early nineteenth century, which achieved efficiencies well in advance of Watt's own engines, Forbes made two points. First he claimed those engines for Watt in the sense that 'they do not contain a single contrivance of any importance ... which was not the unaided invention of Mr. Watt'. Then, immediately, he assimilated Watt and ideas of 'energy': 'In the Cornish engine we see the Energy of Heat rendered available to an extent which the inventor himself [Watt] would at one time have thought scarcely credible'.[14] Note that what Forbes attributes to Watt here is surprise at the amount of energy made available, not surprise at the very concept of energy itself. By so doing his treatment presumes continuity between Watt's conceptual world and the world of 'energy'. The engineer J. A. Ewing, who wrote the new article on Watt for the ninth edition of the *Britannica*, added his own contribution to this tendency by describing Watt as 'having had ideas greatly in advance of his age regarding what is now called energy'. Watt might not have had the term, Ewing thereby claimed, but he had the idea.[15]

All these authors – Cardwell, Joule, Dyer, Forbes and Ewing – were engaged in exploring the terms and conditions on which Watt's heat experiments can be assimilated to the main line of development of thermodynamics. They have different answers to the question and, we might think, varying degrees of success. But they are all engaged in the same sort of exercise, that is, finding a conceptual or procedural reason for attaching Watt to the nineteenth-century energy theorists and thermodynamicists via a line of descent. An alternative way of making this attachment, and one that became increasingly popular during the nineteenth century, was via the indicator and the indicator diagram.

The indicator – the device that we saw placed in Watt's right hand in the commemorative statue placed outside the Watt Memorial Engineering and Navigation School in Greenock – played a central role in this process of assimilating Watt to nineteenth-century thermodynamics. For Watt this was a device with which he thought that he could measure directly the state of the steam in a working engine. In fact, the indicator and the diagrams produced by it are not usually seen in that way because, once again, the historical fog created by that greatest of nineteenth-century scientific developments, the science of thermodynamics, has partially obscured it.

The Indicator Diagram and the Transition from Chemistry to 'Physics'

The origins of the science of thermodynamics (of the relations between heat and motion) as developed by Carnot, Clapeyron, Clausius, Joule, Thomson and Rankine from the 1820s into the 1850s is routinely traced back to the world of the steam engine. This is the example, par excellence, of technology-driven scientific progress, as captured in the famous dictum that 'science owes more to the steam engine than the steam engine does to science'[16]. The 'indicator diagram' is usually treated as a crucial link between Watt's steam engine work and the subse-

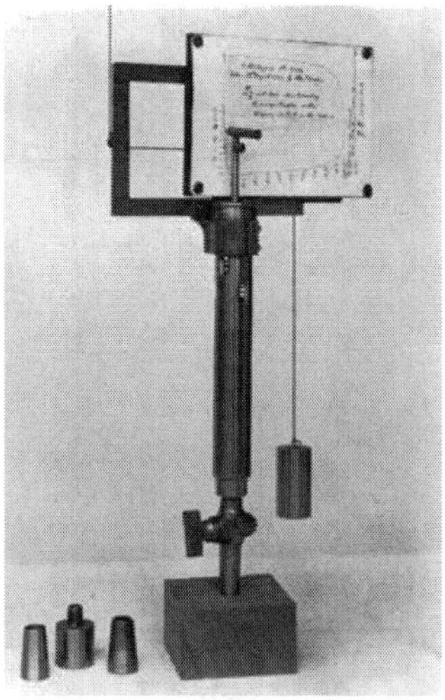

Figure 6.1. Indicator Diagram Apparatus. [Reproduced with permission from Science Museum, London, Science & Society Picture Library]

quent development of thermodynamics. The indicator diagram is thus regarded as a precursor of the insights of that science. Once again, 'precursoritis' proves an obstacle to understanding. I argue that the indicator diagram, in Watt's hands, was not about the physics of heat but rather about diagnosing the chemical characteristics of the steam.

Pictured in Figure 6.1 is an example of the mature indicator diagram technology, dating from about 1840. The apparatus was attached to the cylinder of a steam engine. A small piston moved a pencil on the chart in accordance with the pressure in the engine's main cylinder. The chart itself moved laterally in response to the movement of the piston in the main cylinder and so its 'travel' was in proportion to the working volume of the cylinder at any given time in the cycle. In tracing the origins of the science of energy, the indicator diagram is understandably given an important place. Thus Crosbie Smith says of this device, as developed by Watt and John Southern: 'This 'indicator diagram', the area of which represented the power developed by the engine, allowed for adjustment of the engine in order to maximize useful work. The indicator diagram, more widely known from the 1820s, would subsequently become, in Rankine's terminology, the 'diagram of energy'.[17] For many purposes, this assumption of continuity between the device developed by Southern and Watt and the instrumental and conceptual world of Rankine is useful and unproblematic. But for our current historical purposes it represents a problem, an obstacle to understanding Watt's conceptual world. We need to take Watt out of this 'main line' of development of energy physics, and to try to see the indicator diagram as he first saw it, not as others subsequently reinterpreted it.

In an important sense the indicator diagram was developed because of a pressing technical need. When steam engines were pumping water out of mines, or pumping it for use with water wheels, relatively simple calculations of quantities of water raised through measured distances could be used to provide a measure of power. However, once the rotative steam engine was applied to the powering of machinery, the task became more difficult.[18] Barometric-style pressure gauges were used but these could only supply maximum and minimum pressures in the cylinder and thereby an average pressure. The indicator diagram, as developed by Watt and Southern, in 1796 was also initially conceived as a way of measuring average pressure. But it was subsequently 'realized' that a continuous index of value could be obtained and that the area enclosed by the traced curve was a measure of the power developed by the engine at a given speed (see Figure 6.2).[19] This technique was thus an eminently practical one.

The indicator diagram was something of an 'underground technology' for many years. Although it was used by Boulton & Watt's engineers in building and installing steam engines in a number of countries, it was not widely known. Cardwell claims that it was a well-guarded secret. An incomplete description of it was given in an article in the *Quarterly Journal of Science* in 1822.[20] Watt's description of the indicator, finally published in the same year in Brewster's edition of Robison's *A System of Mechanical Philosophy*, described only the early, non-recording, type of instrument and so left out altogether some essential parts of the mechanism. The British steam expert John Farey apparently did not learn of the indicator until 1819. Some Boulton & Watt men who were erecting a steam engine in Russia had made an indicator, and this came into Farey's possession during a visit to St Petersburg in that year. Farey testified before a Parliamentary Committee on the Patent Laws in 1829 that he had made an indicator on the strength of this and communicated the idea and the process to others, including the author of the paper in the *Quarterly Journal of Science*. This was one, rather roundabout, way that the indicator made its way into the generality of steam-engine practice in Britain.[21] But there were other channels.

The article in the *Quarterly Journal* was dated from Glasgow, 14 February 1822 and signed 'H. H. Jun'. Other clues as to the identity of its author include his statement that the indicator was used in 'our works' and was made from a description provided by 'Mr. Field of London'. It was also observed that the instruments were now made by 'Mr. Hutton of the Anderston Foundry'. On the strength of this information, Dickinson and Jenkins believed that 'H. H. Jun' was probably Mr Hutton or his son. However, it seems much more likely that the mysterious author was in fact Henry Houldsworth Jr (1797–1867).

The Houldsworth family had founded cotton-spinning companies in Glasgow and Manchester, and the Glasgow branch established the Anderston Foundry. Henry Jr became a partner sometime in the 1820s in the Manchester cotton-spinning business founded by his uncle, Thomas Houldsworth, but in the early 1820s Henry junior was still based in Glasgow. He was much interested in innovations in machinery, being implicated in the processes of technology transfer in the textile industry, including by industrial espionage, between the United States and Britain at about that time.[22] Later in his career he was to take out a number of patents. He was accused in the 1830s of unjustly exploiting the ingenuity of his workmen.[23] These aspects of his activities suggest a man who would not rest content for Boulton, Watt and Co. to keep the indicator under wraps. The Houldsworths were no strangers either to Boulton & Watt engines, or to the indicator. Boulton & Watt steam engines were installed in Houldsworth family concerns in both Manchester and Glasgow. One of the earliest indicator diagrams preserved in the Boulton & Watt collection was made in March 1803 at Thomas Houldsworth's mill in Manchester, the diagram showing what gain was made by expansive working and giving information about the valve settings.[24] Correspondence in the Boulton & Watt papers seems to show that various manufacturing operators of steam engines tried to get hold of their own indicators. Whilst the request from Mr Wedgwood to Boulton & Watt for an indicator 'for his own use that he may try his engine under different parts of the machinery', might have been acceded to, it is unlikely that other requests would. As late as 1835 a request to Watt Jr to borrow an indicator was met with the unhelpful reply that Boulton Watt and Co. did not make indicators for sale and that all their instruments were out being used by their engine erectors.[25]

Boulton & Watt employees such as Haden, who erected an engine at Houldsworth's in 1807, used the indicator in setting up and tuning engines and so gave their clients a tantalizing view of the possibilities of the device. But it was unlikely that those clients would have been supplied with a device of their own. While all the details are not clear, it does seem likely that the 1822 article in the *Quarterly Journal* was by Henry Houldsworth Jr and was an outcome of an already longstanding acquaintance of the Houldsworth family with the use of indicators and an equally longstanding frustration of attempts to gain control of one. H. H. Jun's reference to Mr Field as the source of the description of the indicator that his company built is almost certainly a reference to Joshua Field (1781–1863), one of the founders of the Institution of Civil Engineers in 1818, and a long-term partner in Maudsley, Sons and Field. Field was an expert on marine steam engines and in that market he became a competitor of Boulton, Watt and Co. In 1821 Field made a tour of the industrial areas of the Midlands, which included a visit to Soho.[26] Perhaps on that tour he managed to procure details of the indicator that were passed on to Houldsworth and, reinforcing information from Farey, issued via 'H. H. Jun.' in the *Quarterly Journal*.

We should note, as the above account shows us, that considerable technical distance, and quiet industrial contention, separated the indicator of 1785

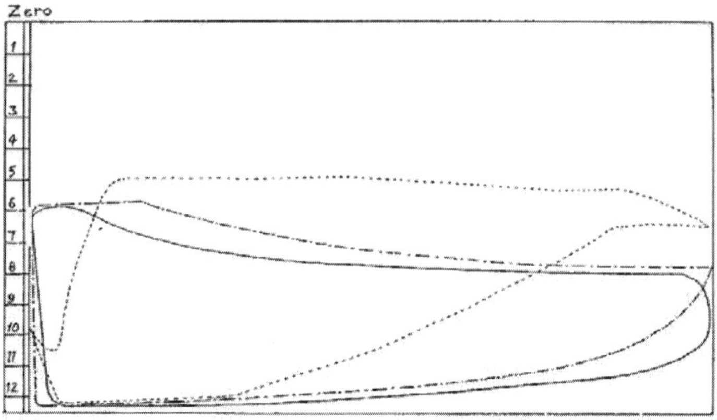

Figure 26 Indicator diagrams taken in 1803 on 'Mr Houldsworth's Engine' in Manchester. The diagrams have been redrawn, the lines on the original being in colour. The significance of the three diagrams shown here is explained as follows:

Dotted line: 'Diagram made with the Valves opening only 1¾ in or less, the Working Gear according to the first construction . . . 19 March 1803.'

Chain line: '[Diagram] made with the Valves opening a *little* in advance. 22d. March 1803.'

Full line: 'Diagram made with the Eccentric Circle advance 15° . . . 25th. March 1803.'

There follows a remark that the difference in area enclosed by the last line and the first 'shews the quantity of steam saved.' Assuming that the engine was running at the same speed as Lee's engine, the area enclosed by the last line shows the power developed to be 9.4 horsepower. (Boulton & Watt Collection.)

Figure 6.2. Indicator Diagram made by Boulton & Watt, 1803 [From Richard L. Hills, *Power from Steam: A History of the Stationary Steam Engine* (Cambridge: Cambridge University Press, 1989), p. 93. Reproduced by permission of Cambridge University Press.]

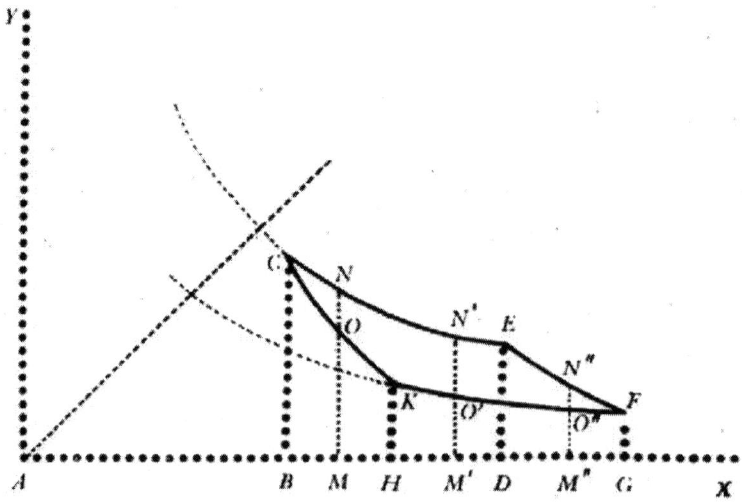

Figure 6.3. '... let us take any gas whatever at temperature T'. Figure resembling the indicator diagram from Émile Clapeyron's 1834 paper. [Reproduced from translation in Eric Mendoza (ed.), *Reflections on the Motive Power of Fire by Sadi Carnot and other Papers... by É. Clapeyron and R. Clausius* (New York: Dover Publications, 1960), p. 75].

from that of 1840 and beyond. Even before knowledge of the indicator became widespread, however, it does appear that Boulton & Watt engineers produced numerous indicator diagrams in the early decades of the nineteenth century that may have come to others' attention. By mid-century the devices themselves and their practical use wherever steam engines were employed were well established. But what is the story of their penetration into the world of engineering science?

In 1834 something looking, as an inscription, very much like the indicator diagram appeared in a paper published by Benoît-Pierre-Emile Clapeyron in the *Journal de l'Ecole Polytechnique* (see Figure 6.3).[27] In the history of thermodynamics, this paper is regarded as having given effective voice for the first time to the key ideas of Sadi Carnot, upon which later figures, notably William Thomson and Rankine would build. In Clapeyron's hands the indicator diagram became an abstract representation of the cycle of a heat engine. There is some circumstantial evidence linking Clapeyron to the indicator diagram as used by Boulton & Watt employees. Clapeyron was in Russia in the 1820s and he may have learned about the indicator diagram there from Boulton & Watt engineers working in St Petersburg.[28] It was certainly whilst in Russia that Clapeyron learned of Carnot's *Réflexions sur la puissance motrice du feu*,[29] and Clapeyron planned his 'Mémoire' whilst in St Petersburg. Cardwell sees the 'Mémoire' as a crucial bridge between the work of Watt and that of the sophisticated theoreticians of a later generation. There is no doubt that later theorists, including William Thomson (Lord Kelvin), were very familiar with the practicalities of Watt's steam engines and, indeed, with the use of the indicator diagram, in its mature form, to measure 'power'.[30]

Although Clapeyron did not mention Watt or the indicator diagram, others made the link explicitly. In later nineteenth-century texts the connection was routinely made. Here, for example, is Peter Guthrie Tait, the close associate and collaborator of William Thomson: 'Besides the many capital improvements which Watt introduced into the steam engine ... we owe to him what is called the *Indicator Diagram*, which is of the utmost importance to the elementary exposition of the fundamental principles of Thermodynamics'. Tait was careful to point out, however, the nature of Watt's understanding of the indicator diagram as giving 'the work done during a stroke of the engine'. It was in the hands of Clapeyron and Rankine that it was developed to represent 'the thermal properties of the working substance itself; and their connection with the laws of thermodynamics'.[31]

The indicator gained important mention when engineers wrote about Watt. One prominent and important example was the article on Watt written for the ninth edition of the *Encyclopaedia Britannica*, by J. A. Ewing.[32] In this first substantial rewriting of the *Britannica* article on Watt since 1824, Ewing has this to say of the indicator:

> One more item in the list of Watt's contribution to the development of the steam-engine is too important to be passed without mention: the indicator, which draws a diagram of the relation of the steam's pressure to its volume as the stroke proceeds, was first used by Boulton and Watt to measure the work done by their engines, and so to give a basis on which the charges levied from their customers were adjusted. It would be difficult to exaggerate the part which this simple little instrument has played in the evolution of the steam-engine. The eminently philosophic notion of an indicator diagram is fundamental in the theory of thermodynamics; the instrument itself is to the steam-engineer what the stethoscope is to the physician, and more, for with it he not only diagnoses the ailments of a faulty machine, whether in one or another of its organs, but gauges its power in health.[33]

By describing the indicator diagram as a 'philosophic notion', Ewing associates Watt with philosophic insight of a thermodynamic kind. Even though he acknowledges the practical purpose of the device for Boulton and Watt, there is no sense that Watt's understanding of the device was a radically different one from that within thermodynamics. Not all engineers readily granted the stethoscope analogy that Ewing accepts here. Rankine, for example, in his *Manual of the Steam Engine*, makes no use of the analogy since he was determined, in line with his beliefs about the intersection of theory and practice, to analyse scientifically what others might accept as the tacit elements of indicator practice.[34]

Why 'Telescoping' the History of the Indicator is a Bad Idea

William Henry Preece gave another, particularly interesting, retrospective account of the indicator in his Watt Anniversary lecture of 1897, which was titled 'Watt and the Measurement of Power'.[35] Preece claimed that the indicator was Watt's 'greatest invention' because it had been 'most prolific in exciting further improvements' in the steam-engine and because it had supplied the engineer with a 'new sense'. Preece, too, echoed the refrain that the indicator was to the engineer what the stethoscope and thermometer were to the physician because it gave an insight into the internal workings of the engine. Crucially, he was also of the opinion that 'few instruments have been found to require less modification' and that Watt's conception and development of it showed that he was 'a long way before his time in grasping the doctrine of energy'. Preece displayed to his audience what he claimed to be Watt's 'original instrument', which he declared to be 112 years old, thus dating it to 1785.[36] He placed this alongside the most recent form of indicator to illustrate their essential similarities and reinforce the point about the virtual completeness of the original device, and thereby support the claim about Watt's prescience. Preece linked the origins of Watt's indicator to his development of expansive working, and specifically to the famous graphical presentation in the 1782 patent. He considered that it was this way of representing what was going on in the cylinder (Preece says 'the work' done in it) that led

Watt to the idea of recording those happenings automatically. Preece remarked on Watt's prescience, considering him 'before his time in the use of graphics, and in the conception of the principle of energy' and thus associated Watt's indicator not only with the concept of 'work' but also that of 'energy'. At another stage, when discussing the 'grand principle of the conservation of energy', he noted that energy is present in a form either useful or useless to man and that the engineer's skill is to render useful that which is useless. Watt more than anyone else, he believed, had helped the engineer to do this.

Preece had borrowed Watt's 'original instrument' from Mr George Tangye of Birmingham. The diagram of it given in Preece's printed lecture shows '1785' inscribed on the main indicator barrel, hence the claim that the instrument was 112 years old. Comparison of the diagram in Preece's paper with the indicator currently held by the Thinktank in Birmingham (see Figure 6.4) shows them to be structurally identical. In fact, a close inspection of that instrument reveals a very faint inscription of '1785' upon the barrel. This shows with reasonable certainty that it is the actual instrument displayed by Preece. This is made even more likely when we learn that the instrument in the Thinktank came from Birmingham Central Library, which had acquired it, along with the Boulton

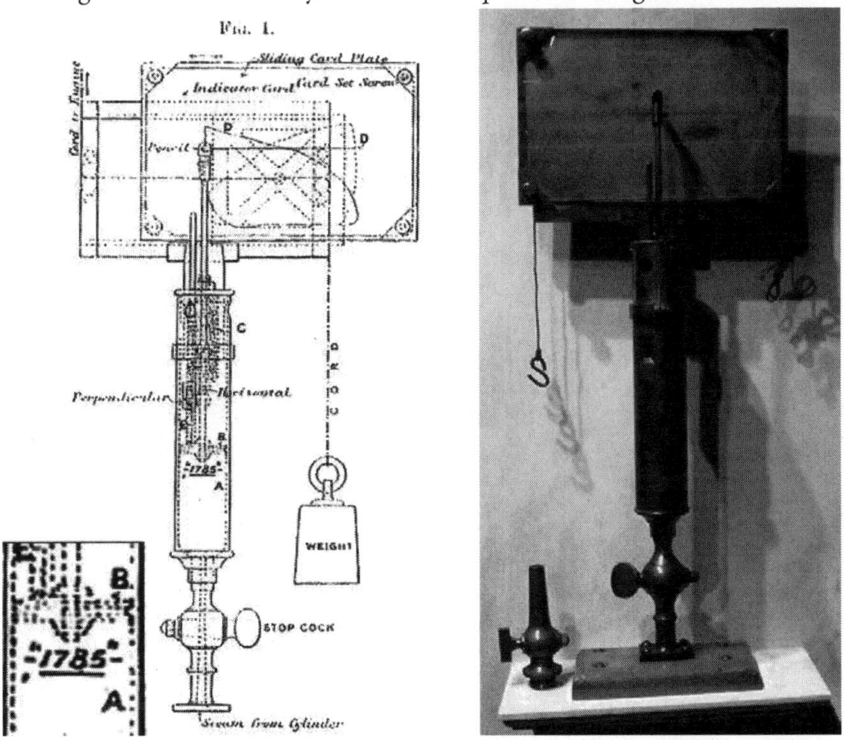

Figure 6.4. Preece's Diagram of Indicator Compared with Indicator in Collection of the Thinktank, Birmingham Science Museum. [Diagram reproduced from W. H. Preece, *Watt and the Measurement of Power. Being the Watt Anniversary Lecture...* (London: William Clowes & Sons, 1897), p. 4. Photograph by permission of Thinktank Trust]

& Watt Collection, from George Tangye in 1911. Tangye himself had bought that Collection in the sale of the Soho Foundry in 1895, which was advertised as including not only the foundry and its machinery but also 'a unique collection of original drawings and atmospheric, sun, planet, and early crank engines, original letters of Bolton and Watt, Telford, Rennie, Arkwright and others, numerous models &c, dating from 1760, of exceptional interest to the engineering world'. The indicator that interests us would have been among this material. The works were sold in one lot on 21 May 1895.[37] So this chain of events would indeed place the 'Thinktank indicator' in Tangye's hands in 1897 when he lent it to Preece to display at the Watt Anniversary Lecture![38]

The problem with all this is that the indicator that Preece displayed as Watt's 'original' device of 1785 is in fact a later model, dating most likely from the 1820s-1840s. An indicator of this design was certainly not extant in 1785 since the first recording indicator dates from the mid-1790s. Preece's erroneous telescoping of the differences between nineteenth-century indicators and Watt's original instrument acted to support his contentions about Watt's theoretical prescience.[39] Revealing the historical problems with Preece's account of the provenance of the instrument, accordingly, renders those contentions problematic. More generally, it may well have been the case that there was widespread ignorance about the history of the indicator in the later nineteenth century that, in similar fashion, brought Watt's 'original' indicator, and therefore the man himself, closer than was warranted to the indicators known and used by the thermodynamicists.

We can grant that there *were* substantial continuities in a physical sense between the indicator diagram as developed and used by Watt and his associates and those who placed similar inscriptions at the centre of their work on the motive power of heat within thermodynamics. But what was the conception of the working of the steam engine that lay behind the indicator diagram for Watt? Given that the 'indicator diagram' evidently plots the continuous variation of pressure in the cylinder during the engine's cycle, it seems natural to assimilate it to later understandings. Although Watt's and Southern's understanding of the device was less fully theorized than that of later investigators, surely the *same thing* was being measured? In fact, the answer to that question has to be 'no'.

The philosopher Davis Baird has already noted the reason for this.[40] Ultimately Baird has a very different purpose from mine for examining the Indicator Diagram, namely to demonstrate the supposed autonomy of the indicator as an instrument across transformations in the theoretical understanding of what it measures:

> What is crucial is that the Indicator displays a phenomenon irrespective of the theories available for understanding that phenomenon. This allowed the Indicator to survive the theoretical ignorance of its birth and promote the theoretical advance

> we now associate with it ... Theories come and go but a new instrumentally-created phenomenon endures. Sometimes such an enduring phenomenon can promote theoretical progress as well.[41]

In pursuit of this account of what he calls 'Thing Knowledge' (about which I have some serious doubts) Baird does, however, present a rather useful description of what he designates (less usefully) as the 'theoretical ignorance' that Watt and Southern brought to the Indicator Diagram, that is, their understanding of what it measured.

As we have seen a number of times already, in his articles on 'Steam' and the 'Steam Engine' in the 1797, third, edition of the *Encyclopaedia Britannica*, Watt's close friend John Robison wrote about steam in terms that were still significantly faithful to the original ideas with which Black and Watt had worked. Reading those articles is to see an earlier cosmology preserved as if in amber. Robison depicted steam as a compound of water with latent heat. Writing about a steam bubble formed in boiling water, he stated:

> Observe, that during its passage up through the water, it is not changed or condensed; for the surrounding water is already so hot that the sensible or uncombined heat in it, is in equilibrio with that in the vapour, and therefore it [the surrounding water] is not disposed to absorb any of that heat which is combined as an ingredient of this vapour, and gives it its elasticity.[42]

So the source of the elasticity of steam is the latent heat. It is the elasticity of steam that enables steam to produce motion by moving a piston to expand the space that it occupies. Thus, if it were possible to increase the latent heat contained in the steam then more motive power would be available. Robison discusses how pressure affects the boiling point of water and therefore the relative content of latent and sensible heat in the steam produced. Davis Baird notices the significance of the fact that then Robison discusses, in the *Britannica* article, a device known as a 'pulse glass'. This is an evacuated vessel containing water or some other substance in which the liquid boils and condenses at temperatures very close to room temperature. (This phenomenon is the basis for novelty 'drinking birds' and the like). In the later republication of Robison's article incorporating annotations by Watt himself, Watt made a revealing comment at the point where Robison discussed the pulse glass. Watt inserted a footnote which read: 'The invention of the pulse glass is ascribed to Dr. Franklin, its date uncertain, probably subsequent to my improvement of the Steam Engine, at least certainly not known to me at that time'.[43] Why does Watt bother to make this statement? As Baird argues, the only reason why Watt would assert that he was ignorant of the pulse glass at the time of his steam-engine improvements would be because he believed that the phenomena that the pulse glass illustrated were relevant to the innovation. Otherwise the denial was unnecessary. This strongly suggests that

Watt *did* believe that water's property of boiling at lower temperatures when at lower pressures was relevant. It was because he believed that the relative proportions of latent and sensible heat in steam affected its elasticity and therefore determined how much motive power the steam could produce in the engine.

Watt and Robison understood the Indicator on this basis. It was originally developed in order to track the details of pressure changes in the cylinder, pressure changes that they believed affected the relative proportions of latent and sensible heat in the steam and therefore its motive power. In another explanatory footnote to Robison's article Watt further described the Indicator as designed to 'show nearly the degree of exhaustion in the cylinder at all periods'. It is important to register that Watt says nothing here about the use of the indicator to determine the work produced by the engine. The focus on the degree of exhaustion of the cylinder is symptomatic of Watt's concern with the way that pressure affected the constitution of the steam, especially the relative amounts of latent and sensible heat within it. It seems likely that over the years Watt had two uses for the indicator, one as an indicator of the elasticity of the steam and the other, later, use as a measure of the work that an engine was performing. Early model indicators in which a pointer simply indicated the pressure of the steam reflected the earlier use. To see this we need to examine in a little detail the early evolution of the indicator.

The earliest, and I think so far unnoticed, record of an attempt by Watt to diagnose experimentally the chemical state of the steam inside a 'working engine' is contained in an eyewitness report by Henry Cavendish. In early August 1785 Cavendish and his assistant Charles Blagden visited the Soho works of Boulton & Watt at Birmingham. They were shown Watt's latest experiments on the steam engine, and Cavendish recorded the experience:

> Mr. Watt mentioned, that having found that some steam is condensed in the cylinder of the Steam engine, tho' surrounded with steam, he made an experiment to discover what happened. He threw steam into a Glass vessel close at the top (By making it communicate with that part of the cylinder of this fire engine in which there is alternately Steam and Vacuum) and found that upon making the Vacuum some of the steam condensed on the sides of the Glass Vessel; and having heated the sides of this Vessel, so that none could condense upon them, he observed the condensation take place, so as to render the Steam visible in the middle of the Glass vessel ... a cloud began to appear at each Vacuum...[44]

The device described by Cavendish is not an indicator in the sense of recording a value or series of values. It was, rather, a way to inspect directly what was going on in the cylinder as the engine went through its cycle. Using this device made available to the experienced eye a sense of the internal workings of the cylinder whilst the engine was in operation. Had Watt been asked to describe what he saw, the account forthcoming would have been in terms of the variable latent heat of the steam. We might call this very first indicator (if such it is) a 'visual' indicator because of its reliance on direct inspection of the state of the steam in the cylinder. It deserves to be designated as an indicator, I think, because of the objective involved in its use, and also because it embodies the idea of attaching a device to an engine which 'communicates' with the cylinder in order to represent in microcosm, as it were, conditions within it.

The first devices to use a surrogate signal to substitute for direct viewing of the steam were simple barometers in which the variation of the level of the mercury indicated the pressure. This sort of barometer had been used to determine pressure in the condenser, but did not work very well for the cylinder. Watt later stated that 'The barometer being adapted only to ascertain the degree of exhaustion in the condenser where its variations were small, the vibrations of the mercury rendered it very difficult, if not impracticable, to ascertain the state of exhaustion of the cylinder at the different periods of the stroke of the engine'.[45] The device usually identified as the first indicator was a pressure gauge attached to the engine cylinder. It consisted of a small cylinder containing a piston controlled by a spiral spring. In this instrument, a pointer was attached to the rod of the piston and indicated on a fixed scale the variation of the pressure in the cylinder. By observing at the same time a reciprocating part of the engine, and noting the pressures, a measure of pressure in the engine at various points in the cycle could be obtained. Examples survive of a modification of this instrument in which a long pointer was attached to the piston rod via a lever (See Figure 6.5). This was the instrument used for the very first attempts to measure the performance of a working steam engine, at the Salford Cotton Mill of George Lee, beginning possibly as early as December 1793.[46] Only at this juncture, in 1796, was the next design of the indicator – the recording indicator – developed, on the initiative of Watt's assistant John Southern. This is the design with the sliding platen, and the pencil attached to the piston arm of the indicator, a design that became familiar and widespread from the 1830s. (See again Figure 6.1) This is the instrument that produces the 'indicator diagram' and the instrument that Preece was misled into putting in Watt's hands in 1785. In truth, in 1785 the visual indicator was all that Watt had. It was to require a decade of development before the recording indicator became available. Only at that stage did it become possible to conceptualize the area of the indicator diagram as a measure of the power of an engine or, more particularly, of the work done by it. Of course, that it was possible to do this at that stage does not in itself mean that it was done.

By means of his early indicators (the 'barometer', the early gauge devices, and those producing early versions of the indicator diagram), Watt was, I suggest, pursuing a 'steam economy' by trying to gauge the state of the steam. This would provide information, for example, about how much cold water to inject into the cylinder. Also, as Robison put it:

Figure 6.5. The Pointer Indicator or Pressure Gauge. [Reproduced from H. W. Dickinson and R. Jenkins, *James Watt and the Steam Engine* (1927), p.229]

> By knowing the internal state of the cylinder in machines of very different goodness, we learn the connection between that state of the steam and the performance of the machine; and it is very possible that the result of a full examination may be, that in situations where fuel is expensive, it may be proper to employ a weak steam which will expend less fuel, although less work is performed by it.[47]

Thus, although a semi-abstract identification of the indicator diagram with the work performed by an engine was beginning to be made at this stage, this was still part of the practical discipline of designing and adjusting engines so as to achieve the optimum steam economy. In that sense it was closer conceptually to Watt's initial attempts to diagnose by literally seeing inside the engine than it was to the nineteenth-century indicator diagram of thermodynamic principle. Indeed Richard Hills notes this in his discussion of the detailed adjustments to Thomas Houldsworth's machine in 1803 using indicator diagrams:

> this indicator diagram shows that, with the final setting of the valves, there was little expansion of the steam during the working stroke and a high terminal pressure so no fall in temperature at the end. *This points to the engine being regarded as working by pressure and not the transformation of heat.*[48]

Once again we see that for Watt at this time, still, heat was what, by chemical means, gave elasticity to the steam and created the pressure that worked the engine. There was no mechanical equivalent of heat here.

The most determined modern argument that a conception of the heat engine did develop within the Watt circle was mounted by Arnold Pacey in 1974. He tried to show, as he put it, that elements of the heat engine concept 'were present as *ideas* before the time of Carnot, and not only as *inventions*'.[49] The key heat engine concept that Pacey claims to find is the idea that there is a flow of heat through the engine. Indeed, in company with Cardwell, he believes that Watt's invention of the separate condenser 'clarified the idea of the *flow of heat through the engine* as an essential feature of its operation'. He relies upon a close examination of experiments with the indicator by the aforementioned George Lee upon his steam engine in the works of Philips and Lee at Salford, near Manchester in 1793–5 to argue that this flow of heat was 'explicitly recognized' by friends and associates of Watt.[50]

This claim is built upon an interpretation of what Lee was doing in four out of a total of fifty-four experiments that were reported to Boulton & Watt in 1796. In those four experiments Lee 'calculated the quantity of heat supplied to the engine and compared it with the quantity of heat given up at the condenser'. The quantities were found to be substantially the same, but Lee would have expected this. As Pacey states, Lee was not testing the conservation of heat but assumed it. His object in making the comparison of output and input was simply to check on the accuracy of the temperature readings. So why all the fuss? What did Lee think of all this? On the basis of evidence that he admits to be fragmentary, Pacey tries to argue that Lee saw heat flowing through the engine and related its flow to the power of the engine:

> Conservation of heat was probably the central principle in Lee's thought (as it still was for Carnot), but Lee must also have had some ideas about how heat gave rise to mechanical power. Indeed, many of his experiments seemed to indicate that the work done was proportional to the quantity of heat given up at the condenser.[51]

But why must he have had such ideas? There is no evidence that he did except in the modernizing gloss that Pacey puts upon Lee's figures by translating the presentation of those figures from, as he puts it, Lee's 'cumbersome form of words' into formulae that Lee did not in fact employ. That Lee measured inputs and outputs of heat says nothing in itself about how he conceived of what was going on in between. Indeed, as Pacey acknowledges 'Lee did not actually calculate power or work, but rather the "effective pressure", that is the pressure difference across the piston. I contend that Lee did this precisely because his idea of how the engine worked had nothing to do with the conversion of heat into mechanical power in the modern sense. Rather, like Watt, he probably conceived of heat combining with water to give elasticity to the steam and so create pressure on the piston. Hills, as we saw above, makes exactly this point about the conceptions guiding the indicator experiments at Houldworth's Mill in 1803, when he concludes that the experiments betrayed the fact that the engine was 'regarded as working by pressure rather than by the transformation of heat'.

Pacey seems intent upon putting words in Lee's mouth and ideas into his head: 'It must have seemed very clear that there was some kind of fixed relationship between heat or temperature and power; but Lee did not try to define it', Pacey states.[52] Similar acts of ventriloquism appear in Pacey's discussion of related work by Lee's and Watt's friend Peter Ewart. In the early nineteenth century a number of publications discussed the relationship between mechanic power or *vis viva*, and the 'force' contained in gunpowder or coal. William Wollaston argued in his Bakerian Lecture before the Royal Society in 1806 that one could be a measure of the other. Others however, were strongly critical of this idea, a writer in the *Edinburgh Review* [John Playfair] contended, on the contrary, that the work done by a steam engine must depend on the way in which the coal was burned, in particular whether it was burned quickly or slowly. This was an idea that Robison also expressed. It was at this stage that Ewart wrote a paper on the concept of 'force' or *vis viva* published a few years later in the *Manchester Memoirs* (1813) in which he defended Wollaston's view. Pacey finds great significance in a quotation from Ewart's paper defending Wollaston. The quotation is as follows:

> however slow or quick the combustion of the coals may be, if they be effectually burnt, the full quantity of heat must be given out. If the heat be allowed to escape without being communicated to the water; or if after being communicated to the water, the pressure of the steam be not wholly applied in producing the intended effect, the loss must be owing to practical imperfections in the construction of the apparatus ... But although the whole heat, or the whole force, can in practice never be completely transferred from one given object to another, yet there can be no doubt of the real existence of both heat and the force in their full quantities...[53]

Of this, Pacey says:

> The idea behind this seems to be of a quantity of heat being transferred from burning coal to the boiler, and then through the engine to the condenser, while at the same time there is a parallel generation of pressure and transfer of 'force' from the cylinder to the machines being driven. The transfer of 'force' must bear a fixed relationship to a particular flow of heat – but no statement by any British experimenter at this time sheds any more light on the problem.[54]

This interpretation of what Ewart is saying is not the only possible one. I see no reason, for example, why the interpretation could not be couched in terms of the chemical combination of heat with water to produce steam. Pacey finds 'the most disappointing feature' of the episode to be that 'Lee and his friends seem to have made very few explicit comments about the significance of the experiments or about the ideas to which they gave rise'.[55] Perhaps their failure to do this resulted from the fact that they saw nothing new or startling in them? Why should we be disappointed at their failure to have more modern ideas?

Pacey, however, continues to seek, and claims to find, evidence that Lee was averse to theorizing, an attitude that might explain his failure to discuss the right things. Writing to James Lawson, a Boulton & Watt employee, Lee gave his response to Prony's second volume of his *Nouvelle Architecture Hydraulique* saying that he was 'all out of patience with Prony ... such a parade of Algebraical Formulae and total want of Candour and Justice to Mr. Watt, and a complete Ignorance of the physical properties of Steam'.[56] Pacey takes this to mean that Lee was an empiricist, averse to theory, and he diagnoses him as having 'a weakness' in 'not seeing the need for clear principles and general concepts on which a theory of steam engines could be built'. On the contrary Lee's words might reasonably be taken to mean that he was happy with Watt's account of the physical properties of steam. On these grounds, I see no compelling reason at all for Pacey's overall conclusion that 'the science of thermodynamics has deeper roots in the eighteenth century than historians of the subject have so far acknowledged'. If it does have such deeper roots, I don't see them as passing through Watt and his immediate circle. It is certainly true, as I argued in Chapter 2, that Watt and his associates were revising some of their conceptions in the early nineteenth century but this was done by way of proofing Watt's claims to philosophic engineering fame against uncertain shifts in the science of heat rather than as an exercise in prescience about thermodynamics.

Conclusion

Norton Wise and other scholars have shown us that indicator diagrams came to have a protean quality in the nineteenth century. Not only have they been seen as crucial to representations within thermodynamics but also modifications of them have been traced in physiology (graphs of muscle and nerve action) and even in the supply and demand curves of political economy.[57] There is no doubt that these connections are real. However, it is also true that they are not continuous or 'natural'. They involved Helmholtz and Fleeming Jenkin in considerable hard work to make them. Wise puts it this way in the case of Helmholtz:

> Conceptually speaking, Helmholtz's adaptation of engine indicators to physiological motive forces is extremely simple. They are of particular interest because they move across immense difference in pressure and time scales. It is easy enough to say that the work of engines replaces that of muscles and therefore both should be subject to the same kind of analysis. It is quite another to compare a fifty-horsepower engine running at fifteen revolutions per minute with a frog muscle developing a few ergs of energy almost instantaneously. Part of Helmholtz's genius, as Olesko and Holmes describe it, lay in his ability to design the extraordinarily sensitive mechanisms that could realize the conceptual analogy, rendering it quite precise in terms of the time development of the muscle's energy.[58]

The point I want to extract from this is that whilst the analogy that moved the indicator diagram from one forum to another was (perhaps) conceptually simple, its realization was complex. In the end it was the intercalation of the theory and detailed practices that sustained the analogical transposition. In exploring analogical moves between thermodynamics/engine practice and physiology or political economy, the 'gap' to be bridged is obvious. It is less obvious, but no less real, I suggest, in the matter of analogical moves between Watt's use and understanding of indicator diagrams and their use by the likes of Clapeyron and Rankine. This was what Tait gestured towards when he compared Watt's practical understanding of the indicator diagram with the 'higher level' to which the thermodynamicists took it. Not only this, I have argued, Watt's practical understanding was underwritten by a theoretical formulation completely alien to the nineteenth century.

Thus those accounts of the Indicator Diagram that assimilate Watt's use of it to its later conceptualization as a measure of work or energy in mid-nineteenth-century thermodynamics are misleading. They read Watt's understanding of the steam engine and the indicator diagram through the historical fog of both the chemical and the thermodynamic revolutions.

7 CONCLUSIONS

James Watt was certainly a 'man o' pairts', but he was also a coherent whole. For too long our historical understanding of him has emphasized the many parts, seeing the whole only dimly, if at all. By pursuing the links between his practical and theoretical work I have tried to set this to rights. Watt was an engineer. He was certainly an expert, as J. D. Forbes put it in his discussions of these questions, at contrivance. Forbes admired Watt's invention of the parallel motion as showing Watt's genius for contrivance, but he argued that this in itself would never have been the basis for 'reputation'. Watt's reputation derived from his quality as a philosophic engineer. I have argued that what gave his engineering its philosophic quality or dimension was largely chemical in character. Watt was a chemist, whose chemistry of heat provided the philosophical dimension to his engineering. That philosophical dimension was very different from the equivalent dimension of what became known in the mid-nineteenth century as 'engineering science'. In engineering science heat was understood as a form of energy, convertible into other forms according to the fundamental laws of thermodynamics. All this was foreign to the intellectual world that Watt inhabited. However, because Watt was adopted as an icon and founding father of their field, the engineering scientists of the nineteenth century were not averse to smoothing over some of the differences between his world and theirs. Watt the chemist and engineer became Watt the mechanical engineer.

In my early chapters I showed how popular representation of Watt, which became a minor industry itself in the nineteenth century, presented the mechanical Watt. This was a natural and easy thing to do given the association of Watt in the popular mind with contrivance. The products of that genius for contrivance, or so it seemed, were everywhere in Victorian Britain, the engines, the linkage and control mechanisms to convert the power of those engines into useful work, the indicators designed to monitor, measure and adjust their performance. Whether as the geometer of Chantrey's statues, reordering mechanical space to institute the separate condenser, or as the bearer of devices, conceived as the product of a craftsman's skill, in later nineteenth-century statues, the mechanical Watt was carved in stone and cast in brass to sit in Westminster, Glasgow, Edinburgh, Birmingham, Manchester, Leeds and Greenock. The indicator that the Watt statue installed in Greenock in the early twentieth century grasped triumphantly in its hand was recognized by those who could recognize it at all either as the stethoscope of the practical working engineer or as the device which produced the 'indicator diagram' that bridged the technological marvels of steam power and the scientific mysteries of thermodynamics. Here, too, Watt the great mechanician was celebrated.

The other ubiquitous image of Watt, one broadcast far and wide by the steam press and new methods of pictorial reproduction developed in the nineteenth century, was the young Watt and the kettle. These images and stories appealed to the idea of youthful genius prefiguring mature accomplishment. Usually the import of the image was nothing more than that. Exactly what was prefigured was rarely of too much concern. The standard presumption was that the young Watt was realizing the power of steam, noticing the fact that steam might be a source of mechanical action. Sometimes the young Watt was seen as contemplating the properties of steam, notably its condensability. But so far as those properties were understood at all by nineteenth-century audiences they were understood in terms of thermal laws – the physics of heat. So this aspect of Watt iconography also tended to reinforce the image of the mechanical Watt.

In textual treatments, the character and status of Watt's achievement was much debated. I have shown that in the late eighteenth century a close connection between Watt's chemistry and his engineering accomplishments could be contemplated. In fact Watt broadcast such a connection, as did others. However, in the nineteenth century that connection was unravelled. The central fact behind the unravelling was the victory of Lavoisian chemistry and the rapid perception of phlogistic chemistry as archaic and misguided. Given this fact, Watt's undoubted engineering achievements and his status as a profound engineer could not comfortably, or reasonably, be connected to his archaic chemical understandings.

This separation of Watt's chemistry and his engineering was pushed along in a number of ways. One was through Watt's self-fashioning in the early years of the nineteenth century. The scientific respectability of Watt's steam experiments, the basis for his already strong reputation as a philosophical engineer, was threatened by the rapidly shifting understandings of heat and the experiments of others. In such a situation the fact of the experiments having been done, and done by Watt as distinct from Joseph Black, had to be publicized. But the substance of them had to be adjusted to the modern idiom, their chemical character being de-emphasized. The account of the experiments finally published in Watt's notes on Robison's essays on 'Steam' and on the 'Steam Engine' in 1822 was, I have suggested, a result of this process. Watt's chemical understanding of heat was assimilated to the thermal laws explored by early nineteenth century investigators.

Another episode acting to alienate chemistry and engineering in Watt's accomplishment was the 'water controversy'. Whereas my earlier study of that controversy emphasized the 'battle of the icons' (Cavendish and Watt) driven by their value to contrasting ideologies of science–technology relations, my concern here has been with its consequences for the depiction of Watt as chemist and as engineer. My conclusion is that neither side in the water controversy could give a true contextual picture of Watt as chemist. Those who denied Watt status as chemical discoverer in the matter of the compound nature of water actively depicted his chemistry as archaic and just plain wrong! However, they had good reasons for wanting to support Watt's status as a philosophical engineer. In order to maintain both these positions, any substantive connection between Watt's chemistry and his engineering (or even an appreciation of how one provided a context for the other) could not be, and was not, countenanced. For their part, Watt's supporters in the controversy sought to sustain his chemical reputation by not venturing at all deeply into the substance of Watt's chemical ideas. Instead they provided a gloss of those ideas as essentially similar to the modernized chemical understandings that had been constructed for Cavendish by his supporters. What was crucial, Watt's advocates argued, was not the detailed substance of Watt's and Cavendish's ideas – both men had approached close enough to the chemical truth about water to have a legitimate claim – but rather the priority in having those ideas and communicating them. They contended that Watt deserved the palm on both grounds. By pursuing this strategy Watt's supporters obfuscated the ways in which the chemistry of heat had informed Watt's steam engine improvements. They continued the work of reinterpretation and re-emphasis that Watt had begun with his construction of the 'Plain Story' of the origins of those improvements.

While early Victorian chemists resisted the adoption of Watt as a hero of their field, the insurgent engineers of the mid-nineteenth century embraced him wholeheartedly. In particular, the engineering elite, who were pushing the status of their field under the banner of 'engineering science' found a useful symbol in the 'Great Steamer'. Not long before, the status of engineer had been very problematic. In fact, I argued that Watt's undoubted self-identification as such was effectively an act of defiance in the learned world of the late eighteenth century, an act designed to assert the status of philosophical engineer against those who saw only lowly tradesmen.

As the engineering profession developed in the nineteenth century, and its major institutions and educational frameworks were established, a number of groups advocated the cause of 'engineering science'. The Gentlemen of Science of the British Association's early years saw 'engineering science' as a way to institute hierarchical control over engineering by physical scientists, and even by gentlemen whose liberal education included mechanical science. Their representations of Watt as a philosophic engineer granted his engineering profound status whilst dismissing his chemistry as superficial. But the Gentlemen of Science were also usually anxious to stress that Watt the autodidact had in fact relied upon Professor Black, and hence upon the academic world, for some of the key ingredients that rendered his engineering philosophic and effective.

The Northern philosophers, such as Thomson, Rankine and Tait, who began from mid-century to eclipse the Gentlemen of Science in control of major scientific institutions, brought a different approach to their advocacy of 'engineering science'. They saw a genuine unification of theory and practice without the clear domination of one by the other. They exemplified in their own backgrounds and careers the easy and untroubled combination of the roles of engineer and natural philosopher. As Peter Bowler and Iwan Morus put it, it was not so much that they 'regarded their researches into the science of heat as being of potential use in improving the design of steam engines but that they regarded the two enterprises – natural philosophical investigation and technological improvement – as being two sides of the same coin'.[1] Their depiction of Watt tended to reflect this attitude. They, unlike the Gentlemen of Science, were comfortable with Watt as a businessman and a patentee, seeing no reason why he could not combine these roles with those of engineer and natural philosopher. Although they were, on occasion, careful to recognize Watt's own intellectual context, their enthusiasm for him as a precursor who personified the unity of theory and practice, sometimes led to elisions that aligned Watt closely, and falsely, with energy physics and thermodynamic insight.

The painting that I have chosen as the frontispiece of this book, 'James Watt and the Steam Engine: Dawn of the Nineteenth Century', which was produced by James Eckford Lauder in 1855, captures marvellously the range of representations of Watt that were constructed in the nineteenth century. Lauder seems to have loaded into his picture all the variety of, sometimes contending, representations of Watt that various interests were generating. We have already seen Rankine's praise of this image for its inclusion of the model Newcomen engine. Some who knew nothing of that model and its supposed significance had criticized the picture for incorporating 'an object so unattractive'. However, Rankine declared the artist right to include it because 'that small and uncouth mass of wood and metal shines with imperishable beauty, as the earliest embodiment of the genius of James Watt'. So Rankine found Lauder's painting inspiring because it showed how the engineer or mechanic could, through engineering science, rise 'to the dignity of a Sage'.[2] The image, however, covers a number of other possibilities. The papers on the desk, with dividers poised, echo Chantrey's Watt as geometrician. Tools arrayed in front of the model engine speak of Watt the craftsman in a gesture of solidarity with the working engineers who found self-affirmation in that image of the great man. Just identifiable behind the model is, it appears, a kettle on a stove. The viewer can pick a favourite symbol and find reinforcement in Lauder's painting for the preferred characterization of Watt. However, the painting, in at least one respect, is perhaps 'loaded' towards a particular view. The positioning of the kettle, consigned as it is to the background to the point of obscurity suggests how in the nineteenth century other symbols obscured the only one likely to lead to a proper appreciation of Watt's approach to the science of steam. The subtitle that Lauder gave to his painting – 'Dawn of the Nineteenth Century' – conveys the idea that Watt's invention caused the sun to rise on a new triumphant era in human material development. To me it also symbolizes the sun setting on a truly contextual understanding of Watt's intellectual world and of his achievement because of the overwhelming urge to render Watt as a symbol of then contemporary utility.

Leaving 'Representations' (except my own) behind, in Part 2 I set out to try to understand the realities of Watt's intellectual universe. I argued that the dominant intellectual construct through most of his scientific and many of his engineering activities was the chemical theory of heat that he imbibed from Cullen and Black, and that he investigated in his kettle experiments. In many respects Watt did more than Black to explore the ramifications of that theory. It was with him throughout his collaborations with Priestley, De Luc and Beddoes and it informed not only his steam experiments and engine improvements but also his work on water, on airs and on pneumatic medicine. In the course of tracing these developments I have tried to show that Watt was not a mere speculative chemist in his collaboration with Priestley but rather he conducted many experiments both in the laboratory and 'in the large'. Particularly significant, I think, but as yet barely glimpsed, are the signs that Watt's thinking about the chemistry of water was partially shaped by observations made of the operation of steam engines. Whether in the laboratory or erecting or testing an engine Watt was ever alert to, and anxious for, chemical understandings. Although Watt was one of the last of the phlogistians, his ideas were not unchanging. His reading as well as his laboratory experiments and his practical ventures, such as chlorine bleaching, meant that Watt routinely imported new ideas into his chemical thinking. We saw from his excursion into the debates about chemical nomenclature that he eventually adopted, and adapted, 'oxygene' as a replacement for De Luc's 'humor' (the principle of 'aquosity'), but did so in a way that preserved, and in his view strengthened, phlogistic chemistry. He was evidently impressed by Berthollet's ideas about affinity and used them in his inquiries into the production and mode of action of medically useful airs. There is, I think, much more to learn about the interactions of Watt's practical and theoretical chemistry, but perhaps these examples point the way. In particular, I hope that future, more detailed, reconstructions of Watt's chemistry in his dealings with the wider world of the British and European Republic of Letters will bear the steam engine in mind rather than shunting it onto a branch line.

It is very difficult for us to think about steam engines as anything other than heat engines in which the steam is merely a 'working substance', a conduit for the passage of heat from one body to another and its conversion into mechanical work. Yet, if we are to think like Watt about steam engines, we have to leave those ideas behind. For him, as I have repeatedly emphasized, steam was a compound of water and elementary heat, and its properties, notably its elasticity, upon which the operation of the cylinder and the viability of the separate condenser depended, derived from this compounding. The traditional arguments about whether Watt's improvements of the steam engine depended on the concept of latent heat or not in many ways miss the point. Certainly, the specific heat of the cylinder and the avoidance of its repeated heating, cooling and reheating were initially more directly important to those improvements than ideas derived from latent heat. But in an important sense the heat latent in steam was a conceptual sub-stratum of *all* Watt's thinking about the steam engine. It surfaced directly in relation to the problem of 'back pressure' and also in his experimental derivation of 'Watt's Law' and conceptualization of the expansive working of engines, but it was always there.

Seeing the steam engine as a chemical device is not just a matter of appreciating the importance of Watt's chemical thinking in his routes to the engine's improvement. It also requires us to consider what I have called the 'ecology of steam'. For Watt's philosophical circle, the chemical theory of heat was important to the understanding of a great variety of natural phenomena, especially geological and meteorological processes. I have shown that there was considerable transposition of understandings from one realm to another. Watt learned from De Luc's meteorology, while the Genevan profited from Watt's steam experiments. Hutton gained much from the chemistry of calcareous materials, of airs and of steam in weaving together his ideas about the role of heat in geological processes. Analogical reasoning abounded as De Luc and Darwin imagined steam engines in the clouds and Hutton conceived the Earth as a heat machine incorporating pistons and spiracles (safety-valves). We must be careful not to use the term 'heat engine', of course, because that disguises the fact that it was the chemical compounding of heat that was important. Apart from their own intrinsic interest and value, these geological and meteorological musings were important reinforcements for, and affirmations of, Watt's understandings of the steam engine. They lent coherence to its place in an understanding of the natural order.

Finally, my analysis moved to the story, and significance of, the indicator and the indicator diagram. The tale behind the placing of this device in the right hand of Watt's statue on the banks of the Clyde is certainly a tangled one. Straightening out the origins and evolution of the indicator technology is important. It shows us, for example, that Watt had begun with the idea of literally seeing inside the steam engine to gauge the state of the steam, to see the physical manifestations of variations in its chemical constitution during the course of a steam cycle. A series of devices culminated, thanks largely to the work of John Southern rather than Watt himself, in the recording indicator. Only at this stage was it possible to make semi-abstract identifications of the area of the indicator diagram with the work done by the engine. This was still a long way conceptually from the 'diagram of energy' of the nineteenth-century thermodynamicists. Partly because of the secrecy that surrounded it for many years the history of the indicator was often 'telescoped'. The case of W. H. Preece's misidentification of a recording indicator as dating from 1785 shows how readily the telescoping of the material history of the indicator could create false impressions of Watt's prescience as an energy physicist and proto-thermodynamicist.

In all these ways, then, the story of the indicator captures much of the process whereby the mechanical Watt eclipsed the chemical Watt in the nineteenth century. The indicator also raises other broader theoretical and historiographical issues going beyond the present study but upon which it throws some light. These issues, which have to do with the whole question of the relations of science and technology, are raised by Davis Baird's notion of 'Thing Knowledge' as applied to the case of the indicator.

While Baird finds knowledge in things and thus collapses science and technology together, I find the continuous inosculation of theory and practice a more viable notion. The way that we conceptualize things is a crucial determinant of what they are. They do not autonomously carry meanings. Of course things do in a sense 'resist' but only by virtue of making it difficult for us to attach meanings to them. Resistance as such is not identifiable. We cannot, then, study Watt purely as an engineer or purely as a natural philosopher. The interaction of the scientific and the practical is what needs to be understood. I have reflected this in my treatment of Watt's chemistry and of his steam-engine improvements. However, this study is predicated on the assumption that it is meaningful to seek analysis of just what the 'science' that entered into the science/engineering amalgam in Watt's ventures was. In Peter M. Jones's terms I treat Watt as 'savant-fabricant' but still feel entitled to ask just what kind of 'savant' he was.[3] I also feel impelled to ask that question. Amalgams of science and practice by historical actors can profoundly mislead us about what earlier amalgams were like. This is why I have felt the need to prise apart the study of heat and the production of the steam engine as a preliminary to putting them back together again in a new, more contextually sensitive, form. James Watt was a chemist. He was also an engineer. We understand him, and the origins of the steam age, better if we can see him as both at the same time.

NOTES

Introduction

1. Roderick Murchison to W. V. Harcourt, 28 December 1839, in J. Morrell and A. Thackray (eds), *Gentlemen of Science. Early Correspondence of the British Association for the Advancement of Science* (London: The Royal Historical Society, 1984), pp. 328–9. Murchison referred to Harcourt's then recent attack on Watt's chemistry as his 'misprision of the Great Steamer'.
2. On Watt's early life see R. L. Hills, *James Watt. Volume 1: His Time in Scotland, 1736–1774* (Ashbourne: Landmark Publishing Ltd, 2002), pp. 29–69, and G. Williamson, *Memorials of the Lineage, Early Life, Education and Development of the Genius of James Watt* (Edinburgh: T. Constable for the Watt Club, 1856).
3. D. J. Bryden, 'James Watt Merchant: The Glasgow Years, 1754–1774', in D. Smith (ed.), *Perceptions of Great Engineers: Fact and Fantasy* (London: Science Museum for the Newcomen Society, National Museums and Galleries on Merseyside and the University of Liverpool, 1994), pp. 23–34.
4. Hills, *James Watt. Volume 1*, pp. 72–3; J. P. Muirhead, *The Life of James Watt with Selections from his Correspondence* (London: John Murray, 1858), pp. 60–73; E. Robinson and D. McKie (eds), *Partners in Science. Letters of James Watt and Joseph Black* (Cambridge MA: Harvard University Press, 1970).
5. For the origins of the story see Muirhead, *Life of Watt*, pp. 82–3; H. W. Dickinson and R. Jenkins, *James Watt and the Steam Engine* (Oxford: Clarendon Press, 1927), p. 23 and n. 3; Hills, *James Watt, Volume 1*, pp. 339–40, 343 n. 71.
6. Hills, *James Watt. Volume 1*, pp. 181–293.
7. On Boulton see H. W. Dickinson, *Matthew Boulton* (Cambridge: Cambridge University Press, 1936); P. M. Jones, *Industrial Enlightenment: Science, Technology and Culture in Birmingham and the West Midlands, 1760–1820* (Manchester: Manchester University Press, 2008). For an unusual and valuable domestic perspective see S. Mason, *The Hardware Man's Daughter: Matthew Boulton and his 'Dear Girl'* (Chichester: Phillimore, 2005).
8. R. E. Schofield, *The Lunar Society of Birmingham: A Social History of Provincial Science and Industry in Eighteenth-Century England* (Oxford: Clarendon Press, 1963). Another excellent account of the group is J. Uglow, *The Lunar Men: The Friends who made the Future, 1730–1810* (London: Faber and Faber, 2002).
9. E. Robinson, 'Matthew Boulton and the Art of Parliamentary Lobbying', *The Historical Journal*, 7 (1964), pp. 209–29.

10. See R. L. Hills, *James Watt. Volume 2: The Years of Toil, 1775–1785* (Ashbourne: Landmark Publishing Ltd, 2005), pp. 161 ff.
11. Watt to William Small, 24 November 1772, in J. P. Muirhead, *The Origin and Progress of the Mechanical Inventions of James Watt*, 3 vols (London: John Murray, 1854), vol. 2, p. 34; D. P. Miller, 'Scales of Justice: Assaying the Reputation of Matthew Boulton and the Partnership of Boulton and Watt', *Midland History*, 34 (2009), pp. 58–76.
12. On Priestley during this period see R. E. Schofield, *The Enlightened Joseph Priestley. A Study of his Life and Work from 1773 to 1804* (University Park, PA: Pennsylvania State University Press, 2004).
13. J. Watt, 'Thoughts on the Constituent Parts of Water and of Dephlogisticated Air; with an Account of some Experiments on that Subject. In a Letter from Mr James Watt, Engineer, to Mr. De Luc, F. R. S.', *Philosophical Transactions of the Royal Society of London*, 74 (1784), pp. 329–53.
14. On Cavendish see the monumental and insightful C. Jungnickel and R. McCormmach, *Cavendish. The Experimental Life* (Lewisburg, PA: Bucknell, 1999) and, on Lavoisier, A. Donovan, *Antoine Lavoisier: Science, Administration, and Revolution* (Oxford: Blackwell, 1993) and the valuable overview of his place in the history of chemistry by M. G. Kim, 'Lavoisier, the Father of Modern Chemistry?', in M. Beretta (ed.), *Lavoisier in Perspective* (Munich: Deutsches Museum, 2005), pp. 167–91.
15. See D. P. Miller, *Discovering Water: James Watt, Henry Cavendish and the Nineteenth-Century 'Water Controversy'* (Aldershot: Ashgate, 2004).
16. R. L. Hills, *Power from Steam: A History of the Stationary Steam Engine* (Cambridge: Cambridge University Press, 1993), pp. 70–2.
17. A. E. Musson and E. Robinson, *Science and Technology in the Industrial Revolution* (Manchester: Manchester University Press, 1969), pp. 251–337.
18. P. M. Jones, 'Living the Enlightenment and the French Revolution: James Watt, Matthew Boulton and their Sons', *The Historical Journal*, 42 (1999), pp. 157–82; E. Robinson, 'An English Jacobin: James Watt, Junior, 1769–1848', *Cambridge Historical Journal*, 11 (1954–5), pp. 349–55. On Watt Jr's 'filial project' see Miller, *Discovering Water*, pp. 83–104.
19. R. L. Hills, *James Watt. Volume 3: Triumph through Adversity, 1785–1819* (Ashbourne: Landmark Publishing Ltd, 2006), pp. 192–224. For Watt's involvement in legal doctrine and courtroom tactics concerning patents see E. Robinson, 'James Watt and the Law of Patents', *Technology & Culture*, 13 (1972), pp. 115–39, and D. P. Miller, 'Watt in Court: Specifying Steam Engines and Classifying Engineers in the Patent Trials of the 1790s', *History of Technology*, 27 (2006), pp. 43–76.
20. D. A. Stansfield and R. G. Stansfield, 'Dr Thomas Beddoes and James Watt: Preparatory Work 1794–6 for the Bristol Pneumatic Institute', *Medical History*, 30 (1986), pp. 276–302; D. P. Miller and T. H. Levere, '"Inhale it and See?" The Collaboration between Thomas Beddoes and James Watt in Pneumatic Medicine', *Ambix*, 55 (2008), pp. 5–28.
21. Dickinson and Jenkins, *Watt and the Steam Engine*, pp. 228–33.
22. Hills, *Watt. Volume 3*, pp. 234–7, 241; H. W. Dickinson, 'The Garret Workshop of James Watt', Science Museum. Technical Pamphlet No. 1 (HMSO, 1929); J. Gibson-Watt, 'James Watt and the Doldowlod Estate. Extracts from the Correspondence and Notes of James Watt and James Watt Jr, 1798–1819', Typescript, Archives, Birmingham Central Library.
23. Muirhead, *Life of Watt*, pp. 501, 509–10.
24. Miller, *Discovering Water*, p. 91.

25. W. A. Smeaton, 'Some Comments on James Watt's Published Account of his Work on Steam and Steam Engines', *Notes and Records of the Royal Society of London*, 26 (1971), pp. 35–42.
26. Muirhead, *Life of Watt*, pp. 432–3, 521, 530. On the various memorials of Watt in his home town see R. M. Smith, *The History of Greenock* (Greenock: Orr, Pollock & Co., 1921), pp. 331–40, and C. MacLeod, *Heroes of Invention: Technology, Liberalism and British Identity 1750–1914* (Cambridge: Cambridge University Press, 2007), pp. 115, 347.
27. F. Jeffrey, 'The Late Mr James Watt', *The Scotsman*, 4 September 1819, p. 5. The text more or less entire is reproduced in Muirhead, *Life of Watt*, pp. 525–30.
28. MacLeod, *Heroes of Invention*, pp. 91–180 and *passim*.

1 Of Statues, Kettles and Indicators: The 'Mechanical' Watt

1. The saga of the peregrinations of the statue between 1960 and 1976 can be followed in papers released in 2007–8 by the U. K. National Archives: British Railways Board, File AN111/638 'The James Watt Statue' and File AN111/639 'James Watt Statue. Clapham'. The statue was given to Heriot-Watt University by St. Paul's in 1996. See Anon, 'Recent Acquisitions in Edinburgh Museums, 1995–97: Supplement', *The Burlington Magazine*, 139 (1997), pp. 580–4, on p. 582.
2. A. Potts, 'Chantrey as the National Sculptor of Early Nineteenth-Century England', *Oxford Art Journal*, 4 (November 1981), pp. 17–27, on p. 20.
3. C. MacLeod, *Heroes of Invention: Technology, Liberalism and British Identity 1750–1914* (Cambridge: Cambridge University Press, 2007), pp. 15–24. On portraiture of the learned see more generally L. Jordanova, *Defining Features: Scientific and Medical Portraits 1660–2000* (London: Reaktion Books/National Portrait Gallery, 2000), and on the instructive example of Newton see P. Fara, *Newton: The Making of Genius* (Basingstoke: Macmillan, 2002) and R. Higgitt, *Recreating Newton: Newtonian Biography and the Making of Nineteenth-Century History of Science* (London: Pickering & Chatto, 2007).
4. The Westminster Abbey website records that the tomb of Sir Lewis Robessart was damaged and that the floor gave way under the weight of the huge Watt statue revealing rows of gilded coffins underneath: http://www. westminster-abbey. org/history-research/ monuments gravestones/people/25784
5. H. Miller, *First Impressions of England and its People* (1856; Boston: Gould and Lincoln, 1860), p. 376.
6. The reference is to F. Arago, *Historical Eloge of James Watt... Translated from the French... by James Patrick Muirhead* (London: John Murray, 1839), pp. 130–1, 183.
7. See, for example, Williamson, *Memorials of the Lineage*, pp. 242–4. Also see Potts, 'Chantrey as the National Sculptor', pp. 20–1.
8. Another famous portrait of Watt (by C. F. Breda) is notable for the geometrical forms in the papers open before its subject.
9. MacLeod, *Heroes of Invention*, pp. 97–108; C. H. Turner (ed.), *Proceedings of the Public Meeting held at the Freemason's Hall on the 18th June 1824, for erecting a Monument to ... James Watt* (London: John Murray, 1824), pp. 8, 10, 13, 36, 43, 75, 84.
10. Reproduced in Muirhead, *The Life of James Watt* (London: John Murray, 1858), p. 523n.

11. See Watt Jr to H. Brougham 13 July 1834 and 14 March 1835, Brougham Papers, University College, London, 27, 513 and 27, 743
12. The suggestion is made in Watt Jr to Brougham, 27 October 1834, Brougham Papers, University College London, 27, 514.
13. It is also interesting to note that the issue of representing Watt's most important improvements graphically was a central one in the patent trials of the 1790s, when his patent specification was attacked because of its failure to include mechanical drawings of a specific contrivance. That omission was a conscious one by Watt and his friends to avoid limiting the scope of the patents to a particular mechanical realization of the (chemical) principles that he sought to claim. To limit scope in that way would be to invite engine pirates to circumvent the patent by producing only slight variations in the specified mechanical form. Chantrey's images of Watt's machinery are in this sense readable as an abstract geometrical account of basic chemical principle, at least so far as the separate condenser is concerned. But of course they were never read as such but rather as a *mechanical* idea of the separate condenser.
14. See T. Wyke, *Public Sculpture in Greater Manchester* (Liverpool: Liverpool University Press, 2005), pp. 117–19 on Theed. On Fehr see A. S. Gray, J. Breach, N. Breach, *Edwardian Architecture: A Biographical Dictionary* (Iowa City, IA: University of Iowa Press, 1986), p. 177.
15. For many years this statue stood in McPhun Park, Glasgow, headless as a result of vandalism. (I will resist the temptation to suggest that the vandals were endorsing the craftsman, rather than the philosopher, view of Watt.) Obviously the head was restored, and the statue was cleaned when the statue was moved to its new site near the 'People's Palace'. It had been commissioned originally in 1864 by the firm of W. & J. Martin, leather merchants, and stood in a niche above the entrance to their works at 8 Baltic St, Bridgeton until it was given to Glasgow Corporation Parks Committee in 1936. ('James Watt Statue for Glasgow Green', *Scotsman*, 17 September 1936, p. 7.)
16. See R. McKenzie, 'Hard Lessons: Public Sculpture and the Education System in Nineteenth-Century Glasgow', in G. Coutts and T. Jokela (eds), *Art, Community and Environment. Educational Perspectives* (Bristol: Intellect Books, 2008), pp. 241–62, on p. 244. Greenshields' statue is now in the foyer of University of Strathclyde's Royal College building. For a picture of the Oxford Museum Watt statue and of the company it keeps see www.oum.ox.ac.uk/pdfs/statues.pdf, and on its history, MacLeod, *Heroes of Invention*, pp. 349, 353.
17. H. Brown, 'Strangest Steamer Launch on the Clyde', *Greenock Telegraph*, 18 March 1988, pp. 1, 15. For further coverage of the travels of the kettle in 1988 see *Greenock Telegraph*, 19 March, p. 11, 16 April, p. 11, 27 April. p. 8. The yellow kettle on Gourock Pier can be seen at http://website. lineone. net/~10th. scouts. gourock/kettle. htm For a while the Inverclyde Amateur Radio Group broadcast from the kettle, which I believe is now no more.
18. I originally located this print at http://www. yokogawa. com/museum/mu_wat_e. htm (accessed 26 May 2003), the website of the 'Museum of measurement' of the Yokogawa Electric Company in Japan, from which it has now been removed.
19. K. Suzuki, '*Yokohama-e* and *Kaika-e* Prints: Japanese Interpretations of Self and Other from 1860 through the 1880s', in H. Hardacre and A. L. Kern (eds), *New Directions in the Study of Meiji Japan* (Leiden: Brill, 1997), pp. 676–87, at. p. 683. The Watt print is reproduced as Figure 44. It is also reproduced in B. Marsden, *Watt's Perfect Engine: Steam and the Age of Invention* (Cambridge: Icon Books, 2002), p. 188.

20. This is taken from a sermon preached on 15 April 2001, Easter Sunday, by the Rev. Lowell E. Grisham, St Paul's Episcopal Church, Fayetteville, Arkansas, reproduced at http://www.stpaulsfay.org/sermon041501.htm (accessed 26 May 2003, now unavailable).
21. See http://www.agencyfaqs.com/advertising/newcamps/766. html and *The Times*, 11 October 1929, p. 6, 22 September 1955, p. 7 and 15 September 1964, p. 15.
22. R. Farson, *Management of the Absurd: Paradoxes in Leadership* (New York: Free Press, 1997). See also R. Thomsett, *Radical Project Management* (Upper Saddle River, NJ: Prentice Hall, 2002).
23. http://www.ibras.dk/montypython/episode38.htm#11 The Presenter is introducing a documentary on medical breakthroughs. He utters some of the truest words ever spoken in jest about the nature of discovery: 'Penguins, yes, penguins. What relevance do penguins have to the furtherance of medical science? Well, strangely enough quite a lot, a major breakthrough, maybe. It was from such an unlikely beginning as an unwanted fungus accidentally growing on a sterile plate that Sir Alexander Fleming gave the world penicillin. James Watt watched an ordinary household kettle boiling and conceived the potentiality of steam power. Would Albert Einstein ever have hit upon the theory of relativity if he hadn't been clever? All these tremendous leaps forward have been taken in the dark. Would Rutherford ever have split the atom if he hadn't tried? Could Marconi have invented the radio if he hadn't by pure chance spent years working at the problem? Are these amazing breakthroughs ever achieved except by years and years of unremitting study? Of course not. What I said earlier about accidental discoveries must have been wrong. Nevertheless scientists believe that these penguins, these comic flightless web-footed little bastards may finally unwittingly help man to fathom the uncharted depths of the human mind'.
24. N. Champion, *James Watt* (Oxford: Heinemann Library, 2001), p. 9.
25. J. P. Muirhead to James Watt Jr, 14 May 1845, Muirhead Papers, Special Collections, University of Glasgow Library, MSGEN 1354/822.
26. Arago, *Historical Eloge of James Watt*, pp. 6–7.
27. James Scott made this engraving in 1849. The art experts whom I have consulted at the National Portrait Gallery and the Courtauld Institute are unaware of the whereabouts of the original painting. This is rather strange given that Robert W. Buss (1804–75) is reasonably well known, if primarily for his abortive role as a stand-in illustrator of Dickens' *Pickwick Papers*. On that saga, see W. Dexter and J. W. T. Ley, *The Origin of Pickwick* (London: Chapman and Hall, 1936).
28. Muirhead to Watt Jr, 14 May 1845, Muirhead Papers, Special Collections, University of Glasgow Library, MSGEN1354/822.
29. Muirhead, *Life of Watt*, pp. 20–1.
30. On Stone see A. L. Baldry, *Marcus Stone, R. A. His Life and Work* (London: Art Journal, 1896). There are in fact two very similar engravings of Stone's painting. The one by James Scott is shown in Figure 1. 7. This was declared in 1869, published by Henry Graves and Co., and reproduced by Baldry. Herbert Bourne produced a second engraving, incorporating specific differences of detail and in a less delicate style. This can be seen reproduced in R. K. Engen (ed. H. Best), *Victorian Engravings* (London: St. Martins, 1975), p. 38. One interesting difference is that Bourne's engraving does not have a sword hanging beside the Watt hearth as Scott's does. Given that the 'action' in the picture would date from the years immediately after 1745, and that after the rising the possession of arms was prohibited by the 1746 Disarming Act, for someone familiar with those aspects of

Scottish history, the sword would have controversial implications concerning the politics of the Watt household. It is interesting that these issues reverberated in Victorian times. I have been unable to locate the original Stone painting and do not know whether or not it included the sword.

31. The only modern reading of this picture that I know of is by Susan Casteras, who confirms that it is a depiction of young male genius. This was a popular genre, with a stream of paintings depicting the early lives of composers, artists, writers, painters and leaders being produced at this time. Casteras is, however, little interested in what the young Watt is up to in the painting except that he is being a youthful, *male* genius. See S. P. Casteras, 'Excluding Women: the Cult of the Male Genius in Victorian Painting', in L. M. Shires (ed.), *Rewriting the Victorians: Theory, History, and the Politics of Gender* (London: Routledge, 1992), pp. 116–46, on 129–32.
32. N. T. Carrington, *The King of the Golden River AND Letting off Steam. Two Plays for Juniors* (London: E. Mathews & Marott, 1936). At one point in the dialogue of *Letting off Steam* James says: 'That's funny. When the steam can't get out of the spout, it pushes up the lid!' Despite its apparent reliance on the Stone painting for inspiration the play thus has Watt perceiving the power of steam rather than its properties. The play also reflects the tendency of late nineteenth and early twentieth-century stories about Watt to couple invocations of his genius with accounts of his hard work. Thus in Scene Two, in an exchange between Watt as an old man and a friend, we read: *'Friend*: You've had a wonderful career! Wonderful luck! *James*: Do you know how you spell the word 'luck'? *Friend*: l-u-c-k. *James*: W-O-R-K'.
33. This collaboration is examined in detail in D. P. Miller, *Discovering Water*, pp. 111–22.
34. James Gibson to James Watt Jr, 8 October 1834, Archives of Soho, MS 3219/6/106, quoted in E. Robinson, 'James Watt and the Tea Kettle: A Myth Justified', *History Today*, 6 (1956), pp. 261–5, on p. 264. The letter is annotated by Watt Jr 'Reports anecdotes of my father's boyhood from Miss Jane Campbell. Copy sent to Mr. Arago 26 Oct^{r}'. It is reproduced in E. Robinson and A. E. Musson, *James Watt and the Steam Revolution* (London: Adams & Dart, 1969), pp. 22–3.
35. A number of kettle pictures, which place the young Watt at the fireside, probably derive from the anecdote conveyed here by Gibson and used by Arago in his spoken éloge.
36. James Watt Jr to J. B. Pentland, 19 January 1835, Archives of Soho, MS 3219/6/106.
37. James Watt Jr to Arago, 22 January 1835, Archives of Soho, MS 3219/6/106.
38. The documents in question had been retained by the late Lord Gibson-Watt when the bulk of the papers kept at Doldowlod House were purchased from him in June 1994 and subsequently housed in the Archives Division of Birmingham Central Library. Only on Lord Gibson-Watt's death did the 'kettle documents', together with books, instruments, furniture and assorted memorabilia come up for sale. See Sotheby's, *The James Watt Sale: Art & Science. The Property of the Late Lord Gibson-Watt and Members of the Gibson-Watt Family* (London: Sotheby's, 2003), pp. 22–3.
39. I am aware of two versions of the memoranda containing the kettle anecdote. The first is that long held at Doldowlod and sold at Sotheby's in 2003. It is a document sent by Miss Jane Campbell to Watt Jr in December 1834. Watt Jr annotated it with the title 'Memoranda of the early Years of M$^{r.}$ Watt, by his cousin, M$^{rs.}$ Marion Campbell (born Muirhead, and daughter of his Mother's sister who was his companion in Youth and intimate friend through life; communicated by her to her daughter Miss Jane Campbell, in 1798, who then wrote them down. Transmitted by her to J Watt in Dec$^{r.}$ 1834'. This document is in Miss Campbell's hand. My claim is that Watt Jr made the change of 'power'

to 'properties' evident in it. The second manuscript version of the document (Muirhead Papers, Birmingham Central Library, MI/8/20, new reference MS3219/6/219) is in a third hand, probably that of someone acting as a copyist for Watt Jr at that time, with annotations by Watt Jr at the end. Watt Jr has also inscribed the date '22 January 1835' at the top of the document. This strongly suggests that it is Watt Jr's retained copy of what he sent to Arago with his letter of that date. This copy differs from the Sotheby's document in several respects: (1) the title is slightly different, ending 'who was his companion in youth & friend through life; dictated to and written down by her daughter Miss Jane Campbell in 1798'; (2) a sentence concerning Watt's carpenter's tools given to him by his father is inserted; (3) the beginning of the quotation concerning the kettle incident has Watt's aunt referring to 'James Watt', rather than 'Jamie Watt'; (4) the document refers to 'properties of steam' although from a difference in the quality of the writing it appears as if this copy may have initially read 'power of steam' and been modified. These differences act as markers in determining which of the two documents was used in subsequent publications. It appears that the second document, as edited by Watt Jr, became the source used by Arago, Watt Jr and Muirhead.

40. Other correspondence with Jane Campbell, also part of lot 5 of the Sotheby's sale, shows Watt Jr correcting related aspects of her account, notably her claim that Watt attended Black's lectures at Glasgow. See Sotheby's, *The James Watt Sale*, p. 22. Watt Jr pointed out that his father had explicitly denied this. We know that Watt Jr was a ruthless modifier of history in the service of his father's reputation. See H. Torrens, 'Jonathan Hornblower (1753–1815) and the Steam Engine: A Historiographic Analysis', in D. Smith (ed.), *Perceptions of Great Engineers: Fact and Fantasy* (London: Science Museum for the Newcomen Society, National Museums and Galleries on Merseyside and University of Liverpool, 1994), pp. 23–34; D. P. Miller, ' 'Puffing Jamie': The Commercial and Ideological Importance of being a 'Philosopher' in the Case of the Reputation of James Watt (1735–1819)', *History of Science*, 38 (2000), pp. 1–21.
41. On Watt's resentment concerning the discovery of the composition of water, and the filial project made of his claims by Watt Jr, see Miller *Discovering Water*, pp. 3–33, 92–104.
42. See A. R. Hall, 'What Did the Industrial Revolution in Britain Owe to Science?', in N. McKendrick (ed.), *Historical Perspectives: Studies in English Thought and Society* (London: Europa, 1974), pp. 129–51, on pp. 140–1.
43. Musson and Robinson, *Science and Technology in the Industrial Revolution*, Eric Robinson's early treatment of the kettle myth as having a genuine historical basis (Robinson, 'James Watt and the Tea Kettle') can plausibly be seen as aiding a general endorsement of Watt's scientific status and therefore a move in these historical debates. See also, M. Jacob and L. Stewart, *Practical Matter: Newton's Science in the Service of Industry and Empire, 1687–1851* (Cambridge, MA: Harvard University Press, 2004).
44. See T. F. Gieryn, *Cultural Boundaries of Science: Credibility on the Line* (Chicago. IL: University of Chicago Press, 1999).
45. S. Smiles, *Lives of Boulton and Watt* (London: John Murray, 1865), p. 89.
46. Compare 'Watt, James', *Chambers's Encyclopaedia*, 10 vols (Edinburgh: W. & R. Chambers, 1860–8), vol. 10, pp. 105–6, and *Chambers's Encyclopaedia*, 10 vols (Edinburgh: W. & R. Chambers, 1895) vol. 10, 1895, pp. 578–9.
47. H. W. Dickinson, *James Watt, Craftsman and Engineer*, (Cambridge: Cambridge University Press, 1936), p. 40. Though painted many years before, James Eckford Lauder, 'James Watt and the Steam Engine: the Dawn of the Nineteenth Century' (see my fron-

tispiece), has some of the flavour about it that Dickinson sought. I argue below that Lauder was trying to be all things to all people so far as Watt symbolism was concerned.

48. Perhaps Watt Jr also read the Buss painting as a depiction of his father contemplating the *properties* of steam. Such a reading is certainly possible even though to us it might seem more 'natural' to read it the other way.
49. Fehr (1867–1940) of London, trained at the Royal Academy Schools from 1885 and was then a studio assistant to Thomas Brock (1847–1922). He exhibited regularly at the Royal Academy from 1887. His best-known work is 'Perseus Rescuing Andromeda' (1893) but he produced numerous public statues of historical figures as well as a number of war memorials. See G. T. Noszlopy and F. Waterhouse, *Public Sculpture of Staffordshire and the Black Country* (Liverpool: Liverpool University Press, 2005), p. 266.
50. *Greenock Telegraph and Clyde Shipping Gazette*, 9 June 1906, p. 4.
51. 'The Watt Statue and Memorial Building', *Greenock Telegraph*, 1 June 1908, p. 2.
52. I can find no discussion of the statue in the Carnegie Papers, even though there is a significant correspondence between Carnegie and the Greenock organizers on other aspects of the Watt Memorial. See, for example, Provost John Anderson to Andrew Carnegie, 5 February 1903, Andrew Carnegie Papers, Library of Congress, Washington DC, Volume 94, ff. 17571–17572; Dr. Caird to Carnegie, 26 August 1903, 7 September 1903, volume 98, ff. 18448–18449 and 18477–18478.
53. Yet another piece of supporting contextual evidence comes from the report in the *Greenock Telegraph* of the unveiling ceremony. That report also recorded a list of models presented to, and purchased for, the Watt Memorial Engineering and Navigation School. Among the models of steamers, steam engines and valves we find models of Watt's indicator and of the McInness–Dobbie external spring indicator. (*Greenock Telegraph*, 1 June 1908, p. 2.)
54. McKenzie, 'Hard Lessons', p. 249.
55. For a brief overview see A. Herléa, 'Indicator', in R. Bud and D. J. Warner (eds), *Instruments of Science: An Historical Encyclopaedia* (London: Taylor & Francis, 1998), pp. 325–8. For a detailed history: J. Walter, *The Engine Indicator: A Short History of the Autographic Patterns from 1800 to the Present Day* (Calgary: The Canadian Museum of Making, 2008), via the website http://www.archivingindustry.com/Indicator/contentback.htm
56. Among those consulted are: J. Bourne, *Handbook of the Steam-Engine: Containing All the Rules Required for the Right Construction and Management of Engines of Every Class* (London: Longman & Co., 1865); J. Hopkinson, *The Working of the Steam Engine Explained by the Use of the Indicator* (London: Simkin, Marshall & Co., 1854); W. B. Le Van, *The Steam Engine and the Indicator: Their Origin and Progressive Development* (London: E. & F. N. Spon, 1889); W. H. King and J. W. King, *Lessons and Practical Notes on Steam, the Steam Engine ... For Young Engineers, Students and Others* (New York: Van Nostrand, 1864), pp. 41–61; R. H. Thurston, *A Handbook of Engine and Boiler Trials and of the Indicator and Prony Brake* (London: E. & F. N. Spon, 1890)
57. Herléa, 'Indicator', attributes the stethoscope analogy to R. H. Thurston in the late nineteenth century, but it is to be found much earlier. An early, and possibly the original, use is in T. J. Main and T. Brown, *The Indicator and Dynamometer with their Practical Application* (London: Herbert, 1847), p. 5.
58. A. Carnegie, *James Watt* (New York: Doubleday, Page & Co., 1905). While Carnegie made £10,000 available he insisted that an attempt be made to raise money from small donations from around the world. To this end an elaborate structure of heads of memo-

rial committees in the US, Canada and Australia as well as in Scotland/England was set up. (See R. Caird to Carnegie, 26 August 1903, vol. 98, ff. 18448–18449 and 7 September 1903, vol. 98 ff. 18477–18478, Andrew Carnegie Papers, Library of Congress, Washington, D. C.) Ten thousand circulars were distributed seeking donations, but a disappointingly small sum was raised. Carnegie sent £5,000 on behalf of 'the American subscribers' (probably most of it came from his own pocket), and subsequently brought the sum up to £10,000. Carnegie's speech at the opening of the School also reflected this view of Watt. See the speech preserved in Andrew Carnegie Papers, volume 253, and reported in *The Greenock Telegraph*, 2 June 1908, p. 4.

59. The account of the unveiling ceremony in the *Greenock Telegraph* tells us that 'Professor Barr' was the official representative of the University of Glasgow. This was Archibald Barr (1855–1931) who became Regius Professor of Civil Engineering and Mechanics at the University in 1889 and through whose efforts the James Watt Engineering Laboratories were established there. See J. W. French, 'Barr, Archibald (1855–1931)', rev. I. F. Russell, *Oxford Dictionary of National Biography*, Oxford University Press, 2004; online edition, Oct 2005 [http://www.oxforddnb.com/view/article/30615]

2 The Demise of the 'Chemical Watt' in the Nineteenth Century

1. Miller, *Discovering Water*.
2. William Irvine (1743–87) graduated MD at University of Glasgow in 1766, where he was a favoured pupil and collaborator of Black. Irvine was lecturer in materia medica at the University from 1766 and lecturer in chemistry from 1769, holding both positions until his early death. Irvine became known for an alternative non-chemical explanation of the phenomenon of latent heat as involving a change in the heat capacity of substances. For the moment we need only note that 'Irvinist' treatments of heat rejected the idea that heat became latent by being chemically combined with other substances. I examine his ideas in more detail in Chapter 4.
3. J. Robison, *A System of Mechanical Philosophy...with Notes by David Brewster*, 4 vols (Edinburgh: John Murray, 1822), vol. 2. This volume is prefaced by a letter from Watt to Brewster on the history of the steam engine. On pp. 1–180 there follow Robison's essays on 'Steam' and the 'Steam-Engine', originally published in the third edition of the *Encyclopaedia Britannica*, as edited and annotated by Watt. Also included is an Appendix by Watt giving further details of his engines and his steam experiments. The key narrative of the history of Watt's initial improvements is in a long note on pp. 113–20.
4. See J. P. Muirhead, *Origin and Progress*, vol. 1, pp. lxxx–xcii. It is also reproduced in Muirhead, *Life of Watt*, pp. 83–91.
5. T. Hodgskin, 'Monument to Mr. Watt', *The Chemist*, No. 16, (26 June, 1824), p. 250. My italics.
6. Anon, 'Mr. James Watt', in *Public Characters of 1802–1803* (London: Richard Phillips, 1803), pp. 526–43, esp. p. 535.
7. G. Chalmers, *An Estimate of the Comparative Strength of Great-Britain, During the Present and Four Preceding Reigns...A new Edition; to which is prefixed, A Dedication to Dr James Currie, the Reputed Author of "Jasper Wilson's Letter"* (London: John Stockdale, 1794), p. xxiii. That Watt was indeed 'the chemist of Birmingham' is supported by what James Currie had written (under the pseudonym of Jaspar Wilson) in the work to which Chalmers was responding: 'The security of property and the liberty of spirit diffused through the nation have called forth the talents of our people ... The genius of

Watt, Wedgewood [*sic*], and Arkwright, has counteracted the expense and folly of the American war'. [See *A Letter, Commercial and Political, Addressed to the Right Hon. William Pitt*, 3rd edn (London: G. G. J. and J. Robinsons, 1793), p. 7]. I thank Christine MacLeod for this reference.

8. It became widely agreed that by later (mid-nineteenth century) standards the Watt patent specification would have been invalid because incomplete. In fact the crucial judgment in relation to James Beaumont Neilson's hot blast patent in the mid-nineteenth century established that the patenting of any one concrete method for realizing a principle effectively gave protection against all other methods of realizing it. The argumentation that supported that decision cited the Watt patent in support of Neilson's case. See A. S. Greenberg, 'Quasi-Patent Rights. A Proposal for the Future', *Journal of the Patent Office Society*, 12 (1930), pp. 13–127, on pp. 21–9. On the hot-blast controversies see R. D. Corrins, 'The Great Hot-Blast Affair', *Industrial Archaeology*, 7 (1970), pp. 233–63.
9. 'Boulton & Watt versus Hornblower & Maberley. Copy from Mr. Gurney's shorthand Notes of the proceedings of the Trial of this case in the Court of Common Please, Guild Hall by a Special Jury before the Right Honourable Lord Chief Justice Eyre. December 16th 1796', Archives of Soho, MS3219/4/227.
10. Muirhead, *Life of Watt*, p. 66.
11. Robison, *A System of Mechanical Philosophy*, vol. 1, p. 148. This remark was in the essay on the 'Steam-Engine', which appeared originally in the *Encyclopaedia Britannica* in 1797.
12. Watt Centenary Committee, James Watt Centenary Commemoration, Sept. 16–20, 1919, *Souvenir Guide Book of Special Exhibit in the Art Gallery, Birmingham* (Birmingham: n. p., 1919), especially 'Plan of Exhibit Gallery, Birmingham Art Gallery', on p. 59.
13. H. W. Dickinson and R. Jenkins, *James Watt and the Steam Engine. The Memorial Volume Prepared for the Committee of the Watt Centenary Commemoration at Birmingham 1919* (Oxford: Clarendon Press, 1927), p. 59 and pp. 361, 367, 370.
14. See C. MacLeod and J. Tann, 'From Engineer to Scientist: Reinventing Invention in the Watt and Faraday Centenaries, 1919–1931', *The British Journal for the History of Science*, 40 (2007), pp. 389–411.
15. The 'Plain Story' is most often referenced to Muirhead, *Life of Watt*, pp. 83–91. It was also printed in Muirhead, *Origin and Progress*, vol 1, pp. lxxxx–xcii. See also R. V. Jones, 'The "Plain Story" of James Watt: The Wilkins Lecture 1969', *Notes and Records of the Royal Society of London*, 24 (1970), pp. 194–220, and W. A Smeaton, 'Some Comments on James Watt's Published Account'.
16. R. Caird, 'James Watt's Contribution to the Advancement of Engineering', in *Papers of the Greenock Philosophical Society* (Greenock: "Telegraph" Printing Works, [1910]). Caird (d. 1915) built most of the vessels of the P&O Company's fleet as well as many other great liners. From 1899–1901 he was President of the Institution of Engineers and Shipbuilders in Scotland. He was also a Fellow of the Royal Society of Edinburgh and an honorary Doctor of Laws of University of Glasgow. He and his firm did much to support the Natural Philosophy Department at the University and he was a member of a committee of Clyde engineers who supported the establishment of the James Watt Engineering Laboratories there. (See *Nature*, 9 December 1915, p. 403.)
17. Caird, 'James Watt's Contribution', p. 5.
18. Torrens, 'Jonathan Hornblower (1753–1815) and the Steam Engine', pp. 23–34.
19. The fact that post-1800 narratives were crafted with reference to earlier ones is, however, indicated by the fact that later statements are found in the archive collected together

with earlier ones, the collection having evidently been done by Watt himself. See especially Archives of Soho, MS3219/4/287, formerly Muirhead IV, Box 18.

20. Private narratives began much earlier. Indeed Watt's famous 1765 notebook of his experiments (of which much more below) is in part a narrative reconstruction of Watt's work. Many modern historians of science recognize that there is no clear demarcation to be made between the production and the communication of scientific knowledge so that primary scientific documents, and even laboratory notebooks, should themselves be seen as narratives presenting and constituting discovery. See for example, F. L. Holmes, 'Scientific Writing and Scientific Discovery', *Isis*, 78 (1987), pp. 220–5, and for an overview J. Golinski, *Making Natural Knowledge: Constructivism and the History of Science* (Chicago, IL: University of Chicago Press, 2005), Chapters 4 and 5.
21. 'An Account of James Watt's Improvements upon the Steam or Fire Engine', Archives of Soho, MS3219/4/287/17.
22. Ibid., pp. 1–2 (My italics).
23. Ibid., p. 4.
24. Ibid., p. 4.
25. Ibid., pp. 4–5. It is notable that the story of Watt's 'eureka moment' on Glasgow Green is not retailed here.
26. 'Copy description engines sent to Mr Pryce Cornwall May 15th 1777', Archives of Soho, MS3219/6/214.
27. On the chemical critique of Desaguliers see A. Donovan, *Philosophical Chemistry in the Scottish Enlightenment. The Doctrines and Discoveries of William Cullen and Joseph Black* (Edinburgh: Edinburgh University Press, 1975), p. 147. Boulton was very pleased with Watt's document. When in London he asked one of their engine erectors, Logan Henderson to send him a copy 'as I want to send a paper on the subject to the greatest Woman in the World, having a very fair opportunity'. This was a reference to Empress Catherine of Russia. (See Boulton to Logan Henderson, 16 August 1777, Archives of Soho, MS3147/3/260.)
28. W. Pryce, *Mineralogia Cornubiensis; A Treatise on Minerals, Mines and Mining* (London, 1778), p. 308 (my italics).
29. Hills, *James Watt, Volume 2*, p. 94.
30. Darwin to Watt, 20 November 1789, in D. King-Hele, (ed.), *The Collected Letters of Erasmus Darwin*, (Cambridge: Cambridge University Press, 2007), pp. 353–4. Darwin added, characteristically, that he wanted 'agreeable...gentlemanlike facts, not abstruse calculations, only fit for philosophers'. For Darwin's treatment of the steam engine in his poem see Chapter 5 below.
31. See Boulton to Darwin, 4 January 1790, partially recounted in King-Hele, *Collected Letters*, pp. 357–8, n1.
32. Watt to Darwin, 31 January 1790, Archives of Soho MS3219/4/287/17 (folder 1). The 'dull history' is appended to this letter.
33. See Miller, 'Watt in Court', pp. 58–9.
34. Muirhead, *Origin and Progress*, vol. 1, p. lxxxi. The dependence of Watt upon Black for the *fact* of latent heat, which is implied here, was evidence for all those later accounts that continued to portray Watt the craftsman engineer as dependent on Black the philosopher.
35. The 'Plain Story' was thus evidently part of a tactic that ran counter to the description of Watt's invention as a chemical one. These were alternative ways of avoiding the identification of Watt's invention as being mechanical and therefore requiring mechanical

specification. Watt and his counsel were, I suggest, opportunistic in moving from one to the other.

36. Watt to Robison, 24 October 1796, reproduced in Robinson and McKie (eds), *Partners in Science*, pp. 232–5, on pp. 233–4.
37. J. Black, *Lectures on the Elements of Chemistry ... Now Published from his Manuscripts by John Robison LL. D.*, 2 vols (Edinburgh: William Creech, 1803), vol. 1, pp. iii–v.
38. Count Rumford [B. Thompson], 'An Inquiry Concerning the Source of the Heat which is Excited by Friction', *Philosophical Transactions of the Royal Society of London*, 88 (1798), pp. 80–102; H. Davy, 'Experimental Essays on Heat, Light, and on the Combinations of Light', in T. Beddoes (ed.), *Contributions to Physical and Medical Knowledge, Principally from the West of England* (Bristol: Biggs & Cottle, 1799), pp. 5–150; C. Smith, *The Science of Energy: A Cultural History of Energy Physics in Victorian Britain* (London: Athlone, 1998), pp. 8–9, 108–9.
39. S. J. Goldfarb, 'Rumford's Theory of Heat: A Reassessment', *The British Journal for the History of Science*, 10 (1977), pp. 25–36.
40. See Black, *Lectures*, vol. 1, pp. 30–5
41. Henry was an old friend of Watt Jr and also knew the elder Watt well, their families having had a long association in connection with bleaching, pneumatic medicine and other matters. See W. V. Farrar, K. R. Farrar and E. L. Scott, 'The Henrys of Manchester. Part I: Thomas Henry (1734–1816)', *Ambix*, 20 (1973), pp. 183–208, on pp. 198, 201, 207; idem, 'Part 2', *Ambix*, 21 (1974), pp. 179–207, on pp. 180–1, 185–6, 198; idem, 'Part 3', *Ambix*, 21 (1974), pp. 208–28; idem, 'Part 4', *Ambix*, 22 (1975), pp. 186–204, on p. 186. On William Henry as a 'convinced calorist' throughout his life, and the reflection of this in various editions of his *Epitome of Chemistry* see W. V. Farrar, K. R. Farrar and E. L. Scott, 'The Henrys of Manchester Part 5', *Ambix*, 23 (1976), pp. 27–52, on p. 41.
42. W. Henry, *An Epitome of Chemistry,* 4th edn (New York: Collins and Perkins, 1808), p. xxiv.
43. See A. Kent, 'William Irvine, M. D.', in idem (ed.), *An Eighteenth Century Lectureship in Chemistry* (Glasgow: Jackson, Son & Company, 1950), pp. 140–50.
44. W. Irvine and W. Irvine Jr, *Essays, Chiefly on Chemical Subjects* (London: J. Mawman, 1805), pp. xxii–xxiv, 7, 47, 119, 456.
45. See reviews of Irvine and Irvine Jr, *Essays* in: *The Monthly Review*, ser. 2, 51 (1806), pp. 257–61, on p. 259; *The Annual Review and History of Literature*, 4 (1805), pp. 878–81.
46. See Chapter 6 below.
47. Drafts of Watt to Laplace, January 1803 and 22 February 1803, Archives of Soho, MS3219/4/287, 7. Significantly, I think, these letters are collected together physically in the archive with the papers documenting Southern's consolidation and repetition of experiments on the elasticity of steam.
48. J. Sharpe, 'An Account of Some Experiments to Ascertain whether the Force of Steam be in Proportion to the Generating Heat', *Memoirs of the Literary and Philosophical Society of Manchester*, Second Series, 2 (1813), pp. 1–14. Sharpe (ca 1768–1834) was a member of the law firm Sharpe, Eccles and Cririe of Manchester. He died in Richmond, Surrey on 28 May 1834. He was elected FRS on 13 April 1826 with John Dalton and William Henry as the lead signatories of his certificate. He was also FSA. See Royal Society of London, Archives, Certificates of Election and Candidature, EC/1826/03, and W. E. A. Axon (ed.), *The Annals of Manchester*, (Manchester: J. Heywood, 1886).
49. J. Dalton, 'Experimental Essays on the Constitution of Mixed Gases; on the Force of Steam or Vapour from Water and in Air; on Evaporation; and on the Expansion of Gases

by Heat', *Memoirs and Proceedings of the Manchester Literary and Philosophical Society*, 5 (1802), pp. 535–602. Sharpe also referred to Robison's article 'Steam' in the 17th volume of the *Encyclopaedia Britannica* and to 'Mr. Betancourt's experiments on the same subject' as other sources of data, but relied preferentially on Dalton's.

50. Sharpe, 'An Account', pp. 3–4.
51. D. S. L. Cardwell, *From Watt to Clausius: The Rise of Thermodynamics in the Early Industrial* Age (Ithaca, NY: Cornell University Press, 1971), pp. 160–1.
52. Anon., 'Article IX. Analyses of Books', *Annals of Philosophy*, 1 (January–June 1813), pp. 459–65, om pp. 459–60.
53. Watt to Peter Ewart, 27 February 1814, extracted in Muirhead, *Origin and Progress*, pp. 346–7.
54. See H. Chang, *Inventing Temperature: Measurement and Scientific Progress*, (Oxford: Oxford University Press, 2004), pp. 168–70. On Dalton and Irvinism see R. Fox, *The Caloric Theory of Gases from Lavoisier to Regnault* (Oxford: Clarendon Press, 1971), pp. 104–15.
55. Robison, *A System of Mechanical Philosophy*, vol. 2, pp. 5–10, on p. 10.
56. See Chapter 4 below.
57. Robison, *A System of Mechanical Philosophy*, vol. 2, p. 16.
58. Ibid., p. 35 note.
59. Ibid., pp. 163, 166.
60. Ibid., pp. 166–8. Southern explained: 'It is well known that if common air be expanded, cold is produced... if a given quantity of it at a given temperature could be gradually expanded, and as it was so expanded, gradually supplied with heat, so as to keep the temperature unaltered, this supply of heat would become *latent*; the thermometer would not show it. It is probable, both from analogy and experiment, that this effect takes place in the expansion of steam. It is *not of this part* of its heat, though latent, and, in the experiments above related, undistinguishable, that I would be understood to speak when I state my opinion to be as just mentioned; but it is of that which, when water alters its state to that of an elastic fluid, becomes essential to it in every degree of elasticity, besides that which belongs to its expanded state. This latter may be called the latent heat of *expansion*, while the other may perhaps properly be called *constitutional*'.
61. Robison, *A System of Mechanical Philosophy*, vol. 2, p. 167 note.
62. These are collected in Archives of Soho, MS3219/6/190 (formerly Muirhead I [MI/7/54]).
63. Ure (1778–1857) is best known as the author of *The Philosophy of Manufactures*, the well-known apologia for the factory system much denigrated by Marx and Engels. Ure is characterised as an uncompromising enthusiast for technological development. He was certainly a promoter of Watt's reputation. See W. V. Farrar, 'Andrew Ure, FRS, and the philosophy of manufactures', *Notes and Records of the Royal Society*, 27 (1972–3), pp. 299–324.
64. A. Ure, 'New Experimental Researches on Some of the Leading Doctrines of Caloric: Particularly on the Relation between the Elasticity, Temperature, and Latent Heat of Different Vapours; And on Thermometric Admeasurement and Capacity', *Philosophical Transactions of the Royal Society of London*, 108 (1818), pp. 338–94.
65. Ibid., p. 339.
66. Ibid., pp. 391–2. Ure refers here to the '*Philosophical Transactions* 84, p. 335', by which he means volume 74 (1784), pp. 329–53, on p. 335 where the quoted words do appear.
67. Ibid., p. 392.

68. Ibid., pp. 392–3.
69. Ure was, however, a controversial figure with a reputation for cantankerousness. His animus against Dalton became well known in the 1820s after he claimed that William Higgins, not Dalton, originated the atomic theory. He was not the most sweetly perfumed ally imaginable. See D. S. L. Cardwell, 'Ure, Andrew (1778–1857)', *Oxford Dictionary of National Biography*, Online edition, Oxford University Press, 2004 [http://www.oxforddnb.com/view/article/28013]
70. A. Ure, *A Dictionary of Chemistry*, 3rd edn (London: Thomas Tegg, 1828), pp. 253, 284–6.
71. Ibid., p. 286.
72. Miller, *Discovering Water*, pp. 59–81, 253–74.
73. Ibid., pp. 17–21. I called the two approaches the 'argument from synonymity and priority' and the 'argument from difference' and regard these as generalizable argumentative strategies in discovery contests. See A. G. Gross, 'Do Disputes over Priority Tell us Anything about Science?' *Science in Context*, 11 (1998), pp. 169–70.
74. Arago, *Historical Eloge of James Watt*. On Arago's agenda see Miller, *Discovering Water*, pp. 106–111.
75. See D. P. Miller, 'Between Hostile Camps: Sir Humphry Davy's Presidency of the Royal Society of London, 1820–27', *The British Journal for the History of Science*, 16 (1983), pp. 1–47; Morrell and Thackray, *Gentlemen of Science*; R. Yeo, *Defining Science: William Whewell, Natural Knowledge, and Public Debate in Early Victorian Britain* (Cambridge: Cambridge University Press, 1993).
76. At his death in 1810, Cavendish was compared to Newton. See Jungnickel and McCormmach, *Cavendish. The Experimental Life*, p. 509. Steven Shapin identified him as a prominent and popular symbol of the link between asceticism and the production of sound knowledge: 'The Philosopher and the Chicken: On the Dietetics of Disembodied Knowledge', in S. Shapin and C. Lawrence (eds), *Science Incarnate: Historical Embodiments of Natural Knowledge* (Chicago, IL: University of Chicago Press, 1998), pp. 21–50, on p. 41. On his reputation among the 'Gentlemen of Science' see Miller, *Discovering Water*, pp. 193–214.
77. W. V. Harcourt, 'Address by the Rev. W. Vernon Harcourt', *Report of the Ninth Meeting of the British Association for the Advancement of Science held at Birmingham in August 1839* (London: John Murray, 1840), pp. 3–69.
78. Harcourt, 'Address', p. 23.
79. Ibid., p. 10.
80. Peacock (1791–1858) was a founding member of the Analytical Society (with Babbage and Herschel) in 1812 and pursued various reform causes at Cambridge for the rest of his life, being a Whig appointment to the Deanery of Ely in 1839. See H. W. Becher, 'Peacock, George (1791–1858)', *Oxford Dictionary of National Biography*, Oxford University Press, Online edition, 2004 [http://www. oxforddnb. com/view/article/21673]
81. [George Peacock], 'Arago and Brougham on Black, Cavendish, Priestley and Watt', *Quarterly Review*, 77 (1845): 105–139, on pp. 114, 131, 133.
82. William Whewell to W. V. Harcourt, 11 February 1840, in *The Harcourt Papers*, vol. xiv, pp. 105–06.
83. W. Whewell, *The Philosophy of the Inductive Sciences*, 2 vols, (London: Parker, 1840), vol. 1, pp. 399–400. Privately Harcourt questioned how Whewell could give as much credit as he did to Lavoisier since, with the role of caloric in his scheme, the Frenchman seemed as guilty as Watt. For Harcourt, only Cavendish escaped the stricture.

84. On the Cavendish Society see W. H. Brock, 'The Society for the Perpetuation of Gmelin: The Cavendish Society, 1846–1872', *Annals of Science*, 35 (1978), pp. 599–617; idem, 'The Cavendish Society's wonderful repertory of chemistry', *Annals of Science*, 47 (1990), pp. 77–80, and Miller, *Discovering Water*, pp. 204–6, 236–7.
85. G. Wilson, *The Life of the Honourable Henry Cavendish* (London: The Cavendish Society, 1851), p. 437.
86. J. D. Forbes, *A Review of the Progress of Mathematical and Physical Science in More Recent Times and Particularly between the Years 1775 and 1850* (Edinburgh: Adam & Charles Black, 1858). This was one of the preliminary dissertations prefixed to the 8th edition of the *Encyclopaedia Britannica*.
87. George Wilson to J. D. Forbes, 17 October [1855], Forbes Papers, University of St. Andrew's Library, Incoming Letters, 1855, no. 128.
88. George Wilson to J. D. Forbes, 15 October 1855, Forbes Papers, University of St. Andrews Library, Incoming Letters, 1855, no. 126.
89. For an extensive survey of coverage of the question in encyclopaedias and chemistry textbooks see Miller, *Discovering Water*, pp. 258–74.
90. This was mainly in nationalistic Scottish literature such as W. Anderson's *The Scottish Nation*, 9 vols, 1882 [entry for Watt in vol. 9, pp. 613–22]. Notable in this regard, but thoroughly unreliable, is Carnegie, *James Watt* which makes the claim for Watt while managing to make no mention at all of Cavendish. An interesting scholarly treatment of the question late in the nineteenth century was T. E. Thorpe's Watt Anniversary Lecture of 1898, published in Thorpe's *Essays in Historical Chemistry* (London: Macmillan & Co. Ltd, 1902) and also in *Nature*, 57 (7 April 1898), pp. 546–51.

3 The 'Mechanical Watt': The Making of a 'Philosophical Engineer'

1. Thus: J. Watt, 'Thoughts on the Constituent Parts of Water and of Dephlogisticated Air with an Account of some Experiments on that Subject. In a Letter from Mr James Watt, Engineer ...'; T. Beddoes and J. Watt, *Considerations on the Medicinal Use, and on the Production of Factitious Airs. Part I by Thomas Beddoes M. D. Part II by James Watt, Engineer. Edition the Third, Corrected and Enlarged* (Bristol: J. Johnson and H. Murray, 1796).
2. See the photograph of the plate in Sotheby's, *The James Watt Sale*, p. 44.
3. See D. P. Miller, 'Connecting Engineering and Natural Knowledge in the Royal Society', paper presented at the '3 Societies' Conference, Keble College, Oxford, July 2008.
4. Ibid., and also D. P. Miller, 'The Usefulness of Natural Philosophy: The Royal Society of London and the Culture of Practical Utility in the later Eighteenth Century', *The British Journal for the History of Science*, 32 (1997), pp. 185–201.
5. Watt's certificate is in Royal Society of London, Archives, Certificates of Election and Candidature, EC/1785/10. Watt was balloted and elected 24 November 1785. 'Inventor' might be thought to face some of the same status difficulties as 'engineer'. However many if not most inventors associated with the Royal Society were gentlemen, and there would have been no automatic connection in Fellows' minds between an inventor and the taint of trade.
6. See Watt to William Small, 24 November 1772, in Muirhead, *Origin and Progress*, vol. 2, p. 34. Watt recorded in his journal having written to Small that 'in recommending me to canals &c, he should observe I would only be engineer not Manager & would have nothing to do with cash or workmen'. ('Journal of James Watt, July 1772–April 1773', Archives of Soho, MS3219/4/134, p. 69.) We must be careful, however, in pushing this perception too far. In some respects Watt was just as astute and effective a businessman

as his partner Boulton, to whom is usually ascribed all the acumen in that department. See Miller, 'Scales of Justice'.

7. 'Letter proposed to be addressed to Mr Prony by J Watt in the year 1803...', in Archives of Soho, MS3219/4/287/7. As we saw in Chapter 2, this theme had been a constant refrain in Watt's earlier written accounts of the history of his inventions as in that distributed to Members of Parliament during the campaign for the extended patent of 1775.
8. See Sotheby's *The James Watt Sale*, p. 102.
9. Maurice Crosland has pointed out to me the 'embarrassment' for my general argument (about Watt as chemist) of the fact that 'in publications he insisted on describing himself as an engineer', and added that it was 'surprising that he felt it necessary to describe his profession like this. It was not as if he was claiming to be a member of an elite society'. (Personal communication, 16 January 2009.) My suggestion that Watt was making a point about the term 'engineer' offers an explanation for this otherwise surprising act of self-identification.
10. This argument is made more fully in Miller, 'Watt in Court'.
11. William Small to Watt, [?] December 1773, extracted in Muirhead, *Origin and Progress*, pp. 71–2, on p. 72. The role of writing in Watt's career was much discussed around this time. Both Small and Boulton encouraged him to either write the treatise on steam or a paper for the *Philosophical Transactions*. Small thought that writing would advance Watt's reputation. When combined with a trip to France it would, as he put it, raise the 'curiosity of the English ... to see the philosophical and travelled engineer', (Watt to Small, 27 October 1773 in ibid., p. 67). However, Small had earlier advised that Watt might still have to reconcile himself to 'engineering in the vulgar manner' – that is civil engineering works – to earn a living. (Small to Watt, 19 October 1771, in ibid., p. 18.) When he heard that Watt was considering going abroad, Small sought to dissuade him by holding up the prospect of a career as a consultant engineer: 'You might live by inventing only an hour in a week for mathematical instrument makers'. (Small to Watt, [?] December 1773, ibid., p. 72.) Boulton also encouraged Watt's writing, but characteristically with a steady eye on its publicity value. See Matthew Boulton to James Watt, 10 July 1776, James Watt Papers MS 3782/12/76/113, reproduced in J. A. Verbruggen, 'The Correspondence of Jan Daniël Huichelbos van Liender (1732–1809) with James Watt (1736–1819) and Boulton & Watt, Supplemented by a Few Related Documents', PhD thesis, University of Twente, 2005, pp. 107–108.
12. On that growth see R. A. Buchanan, *The Engineers: The History of the Engineering Profession in Britain* (London: Jessica Kingsley Publishers, 1989), and for a lively, theoretically informed treatment see B. Marsden and C. Smith, *Engineering Empires: A Cultural History of Technology in Nineteenth-Century Britain* (Basingstoke: Palgrave Macmillan, 2005). See also B. Marsden, 'Engineering' and 'Engineering Science', in J. L. Heilbron (ed.), *The Oxford Companion to the History of Science* (New York: Oxford University Press, 2003), pp. 257–8, and pp. 258–61.
13. On tensions between Boulton & Watt and the organized engineering profession see Miller, *Discovering Water*, pp. 91–2. On collective versus heroic ideology and engineering institutions see MacLeod, *Heroes of Invention*, pp. 268–9.
14. W. Whewell, *The Mechanics of Engineering. Intended for Use in Universities, and in Colleges of Engineers* (London and Cambridge: John W. Parker and J. & J. Deighton, 1841).
15. R. Willis, *Principles of Mechanism, Designed for the Use of Students in the Universities, and for Engineering Students Generally* (London and Cambridge: John W. Parker and J.

& J. Deighton, 1841); B. Marsden, '"The Progeny of these Two 'Fellows'": Robert Willis, William Whewell and the Sciences of Mechanism, Mechanics and Machinery in Early Victorian Britain', *The British Journal for the History of Science*, 37 (2004), pp. 401–34.

16. The following example and quotations are taken from the Airy papers, RGO 6/401, ff. 246 and following, Cambridge University Library.
17. Morrell and Thackray, *Gentlemen of Science*: *Early Years*, pp. 259–60.
18. R. I. Murchison and E. Sabine, 'Address', *Report of the Tenth Meeting of the BAAS Held at Glasgow in August 1840* (1841), pp. xxxv–xlviii on p. xxxv.
19. Morrell and Thackray, *Gentlemen of Science: Early Years*, Figure 23 reproduces the list of toasts on that occasion.
20. Forbes's article was not regarded as a mere popular exposition. Such articles were more important than encyclopaedia articles are today. At this time the best of such articles were regarded as genuine contributions to knowledge, and often were. They were written by experts for other experts, as well as for a broader educated public. This was true even though encyclopaedias were also a focus of much criticism and concern about superficial knowledge. See R. Yeo, *Encyclopaedic Visions: Scientific Dictionaries and Enlightenment Culture* (Cambridge: Cambridge University Press, 2001), pp. 246–76.
21. Forbes, *A Review of the Progress*, p. 3.
22. Ibid., p. 10. Here Forbes cites the work of Carnot, Clapeyron, and Thomson.
23. In Chapter 6, I examine other nineteenth-century accounts of the development of ideas about heat that performed very similar elisions and attributed knowledge of 'thermal laws' to Watt in a way that encouraged the reader to consider Watt a 'modern'.
24. T. B. Macaulay, 'The Literature of Britain: A Speech Delivered at the Opening of the Edinburgh Philosophical Institution on the 4th of November, 1846', in *The Works of Lord Macaulay, complete, edited by Lady Trevelyan*, 8 vols (London: Longmans, Green, and Co., 1866), vol. 8, pp. 377–84.
25. Ibid., p. 381.
26. Editorial, *The Times*, 9 November 1846, p. 4, column B.
27. J. D. Forbes, *The Danger of Superficial Knowledge: An Introductory Lecture to the Course of Natural Philosophy in University of Edinburgh, delivered on the 1st and 2nd of November, 1848* (London: John W. Parker, 1849).
28. Ibid., p. 47.
29. [W. Whewell], 'On Mr. Macaulay's Praise of Superficial Knowledge', *Fraser's Magazine*, 40 (July-Dec 1849), pp. 171–5. For identification of Whewell as the author of this review see Isaac Todhunter, *William Whewell D. D.* (London: Macmillan & Co., 1876), p. 172.
30. Whewell was not always consistent on these issues: see R. Yeo, *Defining Science: William Whewell, Natural Knowledge, and Public Debate in Early Victorian Britain* (Cambridge: Cambridge University Press, 1993), p. 172 and *passim*.
31. Forbes to William Whewell, 19 October 1848, Whewell Papers, Trinity College Library, Cambridge, Add. Ms. a. 204/83.
32. Forbes to William Whewell, 29 October 1848, Whewell Papers, Trinity College Library, Cambridge, Add. Ms. a. 204/84. In this letter Forbes identified Henry, Lord Brougham, among whose many roles was as an advocate of Watt in the 'water controversy', as the 'very beau-ideal of Macaulay's 19th Century man'. (See Miller, *Discovering Water*, p. 256.)
33. See, especially, C. Smith and N. Wise, *Energy and Empire: A Biographical Study of Lord Kelvin* (Cambridge: Cambridge University Press, 1989), pp. 68–75, 497–501; Smith, *The Science of Energy*, pp. 15–30.

34. See Forbes to William Whewell, 9 January 1859, Whewell Papers, Trinity College Library, Cambridge, Add. Ms. a. 204/126.
35. This account of Rankine draws on B. Marsden, 'Rankine (William John) Macquorn (1820–1872)', *Oxford Dictionary of National Biography*, online edition, Oxford University Press, 2004) [http://www.oxforddnb.com/view/article/23133]. See also B. Marsden, 'Engineering Science in Glasgow: Economy, Efficiency and Measurement as Prime Movers in the Differentiation of an Academic Discipline', *The British Journal for the History of Science*, 25 (1992), pp. 319–46; idem, '"A Most Important Trespass": Lewis Gordon and the Glasgow Chair of Civil Engineering and Mechanics, 1840–55', in C. Smith and J. Agar (eds), *Making Space for Science: Territorial Themes in the Shaping of Knowledge* (Basingstoke: Macmillan Press Ltd, 1998), pp. 87–117, and Marsden's biography, *W. J. M. Rankine, 1820–72, and the Making of Engineering Science* (Aldershot: Ashgate, forthcoming).
36. W. J. M. Rankine, *Introductory Lecture on the Harmony of Theory and Practice in Mechanics, delivered to the Class of Civil Engineering and Mechanics in University of Glasgow, on Thursday, January 3, 1856* (London and Glasgow: Richard Griffin and Company, 1856). We need always to remember that the stipulation of the relations of theory and practice was contingent and that any given individual or group was not always consistent in such matters, though if pressed they would make themselves out to be so. See B. Marsden, 'Blowing Hot and Cold: Reports and Retorts on the Status of the Air-Engine as Success or Failure, 1830–1855', *History of Science* 36 (1998), pp. 373–420, on pp. 390–400.
37. Rankine, *Introductory Lecture*, pp. 10–11. Here is the quotation from Macaulay: 'The mathematician can easily demonstrate that a certain power, applied by means of a certain lever, or of a certain system of pulleys, will suffice to raise a certain weight. But his demonstration proceeds on the supposition that the machinery is such as no load will bend or break. If the engineer who has to lift a great mass of real granite by the instrumentality of real timber and real hemp, should absolutely rely on the propositions which he finds in treatises on Dynamics, and should make no allowance for the imperfection of his materials, his whole apparatus of beams, wheels, and ropes, would soon come down in ruin, and with all his geometrical skill, he would be found a far inferior builder to those painted barbarians who, though they never heard of the parallelogram of forces, managed to pile up Stonehenge'. [T. B. Macaulay, *The History of England from the Accession of James the Second* (London: Longman, Brown, Green, Longmans & Roberts, 1856), vol. 3, p. 84]
38. Rankine, *Introductory Lecture*, pp. 11–12.
39. Macaulay, *History of England*, vol. 3, pp. 84–5.
40. For his approach to the indicator see W. J. M. Rankine, *A Manual of the Steam Engine and other Prime Movers*, 3rd edn (1858; London: Charles Griffin and Co., 1866), pp. 334–7, 375–6, 404–5, 422. On the ineradicable role of the tacit see H. M. Collins, *Changing Order: Replication and Induction in Scientific Practice* (Beverly Hills, CA: SAGE Publications, 1985).
41. Rankine, *Introductory Lecture*, p. 22. The historical painting referred to is James Eckford Lauder, 'James Watt and the Steam Engine: The Dawn of the Nineteenth Century' (1855) which features in the foreground a rather impressionistic rendering of the famous model Newcomen engine. See frontispiece. Rankine probably saw the painting in the Royal Scottish Academy's Exhibition in 1855, though it had previously been exhibited

in Glasgow. The possible inspiration of the works of Joseph Wright of Derby was noted at the time. (See 'Royal Scottish Academy's Exhibition', *Scotsman*, 21 March 1855, p. 3.)

42. Rankine, *Manual of the Steam Engine*, p. xxii.
43. A. Barr, *James Watt and the Application of Science to the Mechanical Arts: An Inaugural Address Delivered in the University of Glasgow. November 11th, 1889* (Glasgow: James Maclehose & Sons, 1889), pp. 4–5.
44. Ibid., p. 16.
45. Ibid., pp. 19, 22.
46. J. S. Russell, 'On the Application of the Inventions of Watt to Modern Steam Navigation' [The Watt Anniversary Address for 1867], Papers of the Greenock Philosophical Society – No. 3, in *Publications of the Greenock Philosophical Society* (Greenock: The Society, 1887). On Russell see G. S. Emmerson, *John Scott Russell: A Great Victorian Naval Engineer and Naval Architect* (London: John Murray, 1977); O. Darrigol, *Worlds of Flow: A History of Hydrodynamics from the Bernoullis to Prandtl* (New York: Oxford University Press, 2005), pp. 48–59.
47. Russell, 'Application of the Inventions of Watt', p. 2.
48. Ibid., pp. 2–3.
49. Ibid., p. 3.
50. For all their differences, Russell and Brunel (for whom patent abolitionism was a way of doing business) were at one in their opposition to the patent system. See D. P. Miller, 'Principle, Practice and Persona in the Patent Abolitionism of Isambard Kingdom Brunel', *The British Journal for the History of Science*, 41 (2008), pp. 43–72.
51. See C. MacLeod, 'Concepts of Invention and the Patent Controversy in Britain', in R. Fox (ed.), *Technological Change. Methods and Themes in the History of Technology* (Amsterdam: Harwood Academic Publishers, 1996), pp. 137–53; C. MacLeod, 'James Watt, Heroic Invention and the Idea of the Industrial Revolution', in M. Berg and K. Bruland (eds), *Technological Revolutions in Europe. Historical Perspectives* (Cheltenham: Edward Elgar, 1998), pp. 96–116; MacLeod, *Heroes of Invention*, pp. 249–79, 304. See also C. Pettitt, *Patent Inventions. Intellectual Property and the Victorian Novel* (New York: Oxford University Press, 2004).
52. B. P. Cronin, 'Bramwell, Sir Frederick Joseph, baronet (1818–1903)', *Oxford Dictionary of Scientific Biography*, online edition, Oxford University Press, 2004 [http://www.oxforddnb. com/view/article/32040]
53. Cited from a report in *The Times*, by MacLeod, *Heroes of Invention*, p. 282.
54. Sir F. Bramwell, 'Address by Sir Frederick Bramwell, D. C. L., F. R. S., M. INST. C. E., President', *Report of the Fifty-Eighth Meeting of the British Association for the Advancement of Science, held at Bath in September 1888* (London: John Murray, 1889), p. 5. I take Bramwell's term 'another place' to be a reference to the Royal Society, being an analogical working through of the then common depiction of the BAAS as 'the Commons' and the Royal Society as 'the Lords' in the 'Parliament of Science'.
55. Ibid., p. 4.
56. Ibid., pp. 4–5.
57. See D. E. Nye, *American Technological Sublime* (Cambridge, MA: The MIT Press, 1996).
58. Bramwell, 'Address by Bramwell', p. 23.
59. F. Bramwell, 'Watt, James', *The Dictionary of National Biography*, vol. 60, pp. 51–62.
60. Ibid., pp. 51, 53–4, 62.

61. See T. H. Huxley to Thomson, 2 January 1892; W. H. Preece to Thomson, 1 January 1892; John Phillips to Thomson, 1 January 1892 (from HMS 'Fire Queen', Portsmouth); A. Tait to Thomson, 5 January 1892; J. O. Mitchell to Thomson, 29 February 1892, in Kelvin Papers, Cambridge University Library, Pr 232, 318, 314, 393, and 294.
62. 'Glasgow James Watt Anniversary Dinner', *Scotsman*, 24 January 1898, p. 10.
63. Ibid. The Regnault story was one of Kelvin's favourites. In the published version of Kelvin's 1901 oration on Watt at the Glasgow University Jubilee (discussed below) he says that Regnault told his students that Watt's Law was 'the nearest approach to the truth which he had found among the results of previous experimenters'. See W. Thomson, *Mathematical and Physical Papers*, 6 vols (Cambridge: Cambridge University Press, 1882–1911), vol. 6, pp. 345–362, on p. 357.
64. Kelvin to G. H. Darwin, 20 May 1901, Kelvin Papers, Special Collections, University of Glasgow Library, D127. The draft oration is interspersed between copies of letters in Letter Book LB10 (May-July 1901) from item 18 onwards. (Kelvin Papers, Cambridge University Library, Add 7342, LB10.)
65. William Thomson, Lord Kelvin, 'James Watt: An Oration', in William Thomson, Lord Kelvin, *Mathematical and Physical Papers*, 6 vols (Cambridge: Cambridge University Press, 1882–1911), vol. 6, pp. 345–62.
66. 'Glasgow University Jubilee', *Scotsman*, 14 June 1901, p. 6; Thomson, 'James Watt', p. 359.
67. On the battle between different forms of early motor car see D. Kirsch, *The Electric Vehicle and the Burden of History* (Piscataway, NJ: Rutgers University Press, 2000).
68. 'Glasgow University Jubilee', *Scotsman*, 14 June 1901, p. 6; Thomson, 'James Watt', p. 359.

4 Watt's Chemistry of Heat

1. I do not want to suggest, however, that Watt's practical chemistry was *always* related to his theoretical ideas. Like Black and Cullen before him, whose chemical courses had come to incorporate not only scraps of chemistry relating to *materia medica* but also a panoply of industrially-relevant operations, Watt became an adept 'cookbook' chemist of a strongly empirical type. One of the things that those who defended Black's unwillingness to jump to 'System' were defending against was the idea that the only legitimate chemical operations were those that had been or could be integrated into a unified theoretical system. Watt undoubtedly shared some of this suspicion of the tendency to system.
2. D. P. Miller and T. H. Levere, ' 'Inhale it and See?' The Collaboration between Thomas Beddoes and James watt in Pneumatic Medicine', *Ambix*, 55 (2008), pp. 5–28.
3. It is peculiar, for example, that Watt never participated in the meetings of the Coffee House Philosophical Society in London, the agenda of which overlapped extensively with his interests. He and Boulton were honorary members of the Society. See T. H. Levere and G. L'E. Turner, (eds), *Discussing Chemistry and Steam. The Minutes of a Coffee House Philosophical Society 1780–1787* (Oxford: Oxford University Press, 2002), p. 6.
4. See Hills, *James Watt. Volume 2*, pp. 146–7.
5. P. A. Gerstner, 'James Hutton's Theory of the Earth and his Theory of Matter', *Isis*, 59 (1968), pp. 26–31.
6. A. Donovan, 'James Hutton, Joseph Black and the Chemical Theory of Heat', *Ambix*, 25 (1978), pp. 176–190, on pp. 176–7.

7. In eighteenth-century natural philosophy the term 'principle' was deployed to designate one of the simplest forms of matter from which other substances are formed through combinations with other principles. The term was also used, as in Hutton's case here, to refer to a source of agency. See D. R. Oldroyd, 'The Doctrine of Property Conferring Principles in Chemistry', *Organon*, 12–13 (1976–7), pp. 139–55 and R. Rappoport, 'Rouelle and Stahl – The Phlogistic Revolution in France', *Chymia*, 7 (1961), pp. 73–102.
8. See especially: J. Playfair, *Illustrations of the Huttonian Theory of the Earth* (Edinburgh: W. Creech, 1802); D. R. Oldroyd, *Thinking about the Earth: A History of Ideas in Geology* (London: Athlone, 1996), pp. 86–107; C. W. J. Withers, *Placing the Enlightenment: Thinking Geographically about the Age of Reason* (Chicago, IL: University of Chicago Press, 2007), pp. 127–8. Hutton's concept of 'solar substance', developed in J. Hutton, *Dissertations on Different Subjects in Natural Philosophy* (Edinburgh: Strahan and Cadell, 1792), was a crucial cause of embarrassment for Playfair. For Hutton, however, 'solar substance' was a central aspect of his theory, not least because it unified chemical change and the phenomenon of light. An important criticism of the new chemistry in the 1790s and into the 1800s was that it failed to account for such phenomena. This was argued by the young Henry Brougham, surveying the chemical scene for the *Edinburgh Review* in the course of a review of Robison's edition of Black's lectures: H. Brougham, 'Lectures on the Elements of Chemistry', *Edinburgh Review*, 2 (1803), pp. 1–26.
9. Robinson and McKie (eds), *Partners in Science*, p. 3.
10. I. Inkster, 'Thoughtful Doing and Early Modern Oeconomy', in L. Roberts, S. Schaffer and P. Dear (eds), *The Mindful Hand: Inquiry and Invention from the Late Renaissance to Early Industrialisation* (Amsterdam: Koninklijke Nederlandse Akademie van Wetenschappen, 2007), pp. 443–52, on p. 446.
11. Donovan, *Philosophical Chemistry*, p. 182.
12. Robert Anderson makes the point that most of Black's students, unlike Boerhaave's, would not become physicians. See R. G. W. Anderson, 'Boerhaave to Black: The Evolution of Chemistry Teaching', *Ambix*, 53 (2006), pp. 237–54.
13. Black's lectures show that his other chemical ideas interacted very little with his chemistry of heat as such. In many ways Black's long-term reputation was helped by this since it made it relatively easy to prize apart his achievements into chemical discovery (e. g. of 'fixed air') and physical discovery (the 'thermal laws' relating to latent heat).
14. See J. Golinski, *Science as Public Culture: Chemistry and Enlightenment in Britain, 1760–1820* (Cambridge: Cambridge University Press, 1992), pp. 38–9. Golinski gives important insights into the way that the different social contexts in which they operated gave Black and Watt rather different attitudes towards, and possibilities for, the conduct of schemes to apply knowledge as 'projectors' and the claiming of priority in discovery (pp. 43–9).
15. Hills, *James Watt, Volume 1*, pp. 75, 80,156. Black also lent Watt substantial sums of money, the debts not being finally cleared until 1784.
16. Ibid., p. 80.
17. Watt to Brewster, May 1814, printed in Robison, *System of Mechanical Philosophy*, vol. 1, pp. iii–x, on p. ix.
18. W. Cullen, 'Of the Cold Produced by Evaporating Fluids, and some other Means of Producing Cold', in *Essays and Observations. Volume II* (Edinburgh: G. Hamilton and J. Balfour, 1756). On the continuities, and discontinuities, between the chemical teaching of Boerhaave, Cullen and Black see R. G. W. Anderson, *The Playfair Collection and the*

Teaching of Chemistry at the University of Edinburgh 1713–1858 (Edinburgh: The Royal Scottish Museum, 1978), Chapters 1 and 2.

19. Hills, *James Watt, Volume 1*, p. 143; Cardwell, *From Watt to Clausius*, pp. 34–5.
20. R. G. W. Anderson,'Joseph Black and his Chemical Furnace', in idem, J. A. Bennett and W. F. Ryan, *Making Instruments Count: Essays on Historical Scientific Instruments Presented to Gerard L'Estrange Turner* (Aldershot: Ashgate Variorum, 1993), pp. 118–26, on p. 120. See also Anderson, *The Playfair Collection*.
21. Cardwell, *From Watt to Clausius*, p. 35.
22. Hills, *James Watt, Volume 1*, p. 304; Donovan, *Philosophical Chemistry*, p. 221.
23. J. Golinski, ' 'Fit Instruments': Thermometers in Eighteenth-Century Chemistry', in F. L. Holmes and T. H. Levere (eds), *Instruments and Experimentation in the History of Chemistry* (Cambridge, MA: The MIT Press, 2000), pp. 185–210, on p. 197. Donovan, *Philosophical Chemistry* argues convincingly that Black's early investigations of causticity and of fixed air were 'guided by an underlying conception of the general laws which link heat, elective attraction and change of state'. (p. 215). Thus, although Black did not do much by way of sequel to pursue the other chemical consequences of the concept of latent heat, he arguably did so in this 'prequel'.
24. Golinski, ' 'Fit Instruments' ', p. 197.
25. On the contrast between Robison's private and published views on Black and their lasting consequences for historical treatments of him see the important paper by J. R. R. Christie, 'Joseph Black and John Robison', in A. D. C. Simpson (ed.), *Joseph Black 1728 -1799. A Commemorative Symposium*, (Edinburgh: Royal Scottish Museum, 1982), pp. 47–52. See also, Brougham, 'Lectures', and J. B. Morrell, 'Professors Robison and Playfair and the *Theophobia Gallica*: Natural Philosophy, Religion and Politics in Edinburgh, 1789–1815', *Notes and Records of the Royal Society of London*, 26 (1971), pp. 43–63.
26. Fox, *The Caloric Theory of Gases*, pp. 104–20, 154–6.
27. Kent, 'William Irvine, M. D. ', p. 144.
28. Watt wrote to William Small from Glasgow on 6 May 1774 that 'My friend Dr. Irvine in winter discovered what is the lowest possible degree of heat, or the real beginning of the scale; by which means thermometers may be made showing the real quantity of heat in bodies'. (See Muirhead, *Origin and Progress*, vol. 2, p. 78.)
29. Adair Crawford studied with Irvine at Glasgow and applied his teacher's doctrines to animal heat. His book *Experiments and Observations on Animal Heat* (London: J. Murray, 1779; 2nd edn, 1788) enjoyed wide circulation and its ambiguities led to Crawford being given credit for many of Irvine's ideas, and some of Black's. Watt was incensed by this, and noted that Crawford 'seems not to have been much disposed to do either Dr. Black or Dr. Irvine justice'. Watt believed that the only valuable parts of Crawford's voluminous writings were those that he had purloined from Black and Irvine. He thought that Crawford's own experiments and theories were 'of little value, although they serve as one of the props to the French theories'. (See Watt to Adam Ferguson, 8 February 1801, extracted in Muirhead, *Origins and Progress*, vol. 2, pp. 273–5, on p. 275.)
30. For the experiments recorded by Irvine see 'Journal of James Watt, May 1773 – March 1774', Archives of Soho, MS3219/4/135, p. 107.
31. Much attention has been lavished, of course, on the relationship between Black's discovery of latent heat and Watt's improvements of the steam engine, on which see Chapter 5.
32. Lecture, 17 November, 1766 in Blagden MS 'Notes of Dr Black's Lectures', 1766–7, Wellcome Institute for History of Medicine, as quoted in H. Guerlac, 'Joseph Black's

Work on Heat', in A. D. C. Simpson (ed.), *Joseph Black 1728 -1799. A Commemorative Symposium* (Edinburgh: Royal Scottish Museum, 1982), pp. 13–22, on p. 13.

33. This section relies on Guerlac, 'Joseph Black's Work on Heat'.
34. Black, *Lectures on the Elements of Chemistry*, vol. 1, pp. xliii–xliv, 238 and vol. 2 pp. 237–8.
35. Black to Watt, 19 February 1768 and Watt to Black, 7 April 1768, in Robinson and McKie (eds), *Partners in Science*, pp. 10–12.
36. Watt to Black, 24 July 1779 in Robinson and McKie, *Partners in Science*, pp. 65–6, on p. 66. The reference is to P. J. Macquer, *Dictionnaire de Chymie* (Paris, 1766; 2nd edn, revised and enlarged, 1778). I do not know who could have been 'teasing' Watt in this way. It is usually considered that advocacy of mechanical theories of heat was quite rare during this period.
37. Watt to Black, 25 September 1783, in Robinson and McKie, *Partners in Science*, p. 128.
38. D. R. Dyck, 'The Nature of Heat and its Relationship to Chemistry in the Eighteenth Century', Unpublished PhD thesis, University of Wisconsin, 1967.
39. The idea is expressed in J. T. Eller, 'Seconde dissertation sur les elemens', *Mémoires de l'Academie Royal des Sciences et Belles Lettres*, [Berlin], 2 (1746), pp. 25–48, on 43–4. This is held at Glasgow University Library Special Collections, but that copy was acquired as part of the Hunterian Collection in the early nineteenth century. If Watt was able to consult Eller's work while at Glasgow then it would probably have been in the library of an individual whom he knew, such as Black or Anderson.
40. See H. Metzger, *Newton, Stahl, Boerhaave et la Doctrine Chimique*, (Paris: Félix Alcan, 1930).
41. Hills, *James Watt. Volume 1*, p. 143. As late as 1769 Watt needed help in reading German. (Hills, *James Watt. Volume 1*, p. 148.) Much of Eller's work, however, including the key memoir noted above, was published in French in which Watt was reasonably proficient.
42. The manuscript is published, with extensive, valuable background and explication in R. McCormmach, *Speculative Truth: Henry Cavendish, Natural Philosophy, and the Rise of Modern Theoretical Science* (Oxford: Oxford University Press, 2004).
43. McCormmach, *Speculative Truth*, p. 87. The first version of the material theory was held by Black and by Watt and also by William Cleghorn in his inaugural dissertation at Edinburgh in 1779. See D. McKie and N. H. de V. Heathcote (eds, trans) 'William Cleghorn's *De Igne* (1779)', *Annals of Science*, 14 (1958), pp. 1–82.
44. Robison, 'Steam', p. 734; Robison to Watt, 25 February 1800, in Robinson and McKie (eds), *Partners in Science*, pp. 335–41 on p. 337.
45. Quoted in Muirhead, *Origin and Progress*, vol. 2, p. 167.
46. James Watt to Joseph Black, 13 December, 1782, transcribed in Robinson and McKie, (eds), *Partners in Science*, pp. 117–19.
47. Ibid., on pp. 117–118.
48. Watt, 'Thoughts on the Constituent Parts of Water and of Dephlogisticated Air'. See detailed discussion of this paper below.
49. See Robison to Watt, 18 December 1799 in Robinson and McKie (eds), *Partners in Science*, pp. 320–6.
50. Robison to George Black Jr, 21 December 1799 in Robinson and McKie, *Partners in Science*, pp. 326–7 on p. 327.
51. Robison to Watt, 25 February 1800 and Robison to Watt, 23 July 1800 in Robinson and McKie (eds), *Partners in Science*, pp. 335–41 (on p. 339) and pp. 342–7 (on pp. 343–5).

52. Robison to George Black Junior, 18 October 1800 in Robinson and McKie (eds), *Partners in Science*, pp. 360–3 on p. 361. On the avoidance of system as methodological precept see Henry Brougham's friendly review of Black's *Lectures* in *Edinburgh Review*, 2 (1803), pp. 1–26. See also G. N. Cantor, 'Henry Brougham and the Scottish Methodological Tradition', *Studies in the History and Philosophy of Science*, 2 (1971), pp. 69–89.
53. Robison to Gilbert Hamilton, 14 August 1803 in Robinson and McKie (eds), *Partners in Science*, pp. 378–80 on p. 379.
54. Boulton to Watt, 26 July 1781, 28 July 1781, Archives of Soho, MS3147/3/5. See also Schofield, *The Enlightened Joseph Priestley*, pp. 160–1 where Schofield puts this episode into the broader context of Priestley's willingness at this time to put his skills at the service of the practical ventures pursued by his new Lunar Society associates. On the subsequent, mainly nineteenth century, history of air engines see Marsden, 'Blowing Hot and Cold'.
55. See Schofield, *The Lunar Society of Birmingham*, p. 189; D. King-Hele (ed.), *The Collected Letters of Erasmus Darwin* (Cambridge: Cambridge University Press, 2007), p. 204, letter 82(4), note 5.
56. Schofield, *Enlightened Joseph Priestley*, p. 150.
57. King-Hele (ed.), *Collected Letters of Darwin*, p. 182, note 2 and p. 204, note 5.
58. 'Memoirs of Dr. Joseph Priestley (written by himself)', reproduced in *Autobiography of Joseph Priestley* (Bath: Adams & Dart, 1970), pp. 69–139 on p. 120.
59. Priestley to James Keir, 4 February 1778, in R. E. Schofield (ed.), *A Scientific Autobiography of Joseph Priestley (1733–1804). Selected Scientific Correspondence Edited with Commentary* (Cambridge, MA: The M. I. T. Press, 1966, pp. 163–4.
60. Boulton to Watt, 21 July 1781, Archives of Soho, MS3147/3/5. If Watt knew of Boulton's assistance to Priestley with his laboratory apparatus (as seems likely) then the obligation of which Boulton writes must have involved some other form of support, patronage or assistance to Priestley that we, as well as Watt, are unaware of.
61. Schofield, *Enlightened Joseph Priestley*, p. 161; Schofield, *Lunar Society*, pp. 201–2, 333, 241; Watt to Priestley, n. d. 'Thursday', Royal Society, Box PH 13, cited in Hills, *James Watt, Volume 2*, p. 218.
62. J. Priestley, *Experiments and Observations Relating to Various Branches of Natural Philosophy, with a Continuation of the Observations on Air II* (Birmingham: J. Johnson, 1781), p. 388; Schofield, *Enlightened Joseph Priestley*, pp. 153–4.
63. Priestley, *Experiments and Observations on Natural Philosophy*, vol. 2, p. 317.
64. Watt, 'Thoughts on the Constituent Parts of Water'. Watt also published a sequel, rarely noticed: 'Sequel to the Thoughts on the Constituent Parts of Water and Dephlogisticated Air. In a Subsequent Letter from Mr. James Watt, Engineer, to Mr. De Luc, F. R. S. ', *Philosophical Transactions of the Royal Society of London*, 74 (1784), pp. 354–7.
65. De Luc (1727–1817) was a prominent Genovese citizen who, from his appointment as Reader to Queen Charlotte in 1775, became a significant figure in the Royal Court at Windsor. From the early 1780s Watt and De Luc developed a close relationship involving frequent exchange of natural philosophical news and views. As we will see, they occasionally conducted research together. De Luc partially oversaw the Continental educational tourism of James Watt Jr. An anecdotal, but useful account of De Luc is P. A. Tunbridge, 'Jean André De Luc, F. R. S. (1727–1817)', *Notes and Records of the Royal Society of London*, 26 (1971), pp. 15–33. A reputation for having plagiarized Joseph Black, together with De Luc's scriptural geology – much criticized by Charles Gillispie in *Genesis and Geology* – have perhaps too long delayed the serious assessment of his

contributions and importance by historians of science. See M. J. S. Rudwick, 'Jean-André de Luc and Nature's Chronology', in C. L. E. Lewis and S. J. Knell (eds), *The Age of the Earth from 4004BC to AD2002* (London: The Geological Society, 2001), pp. 51–60; D. R. Oldroyd, 'Jean-André de Luc (1727–1817) and the Relationship between Geology and Religion: An Atheist's Comparative View of Historiography' (forthcoming)

66. Watt, 'Thoughts on the Constituent Parts of Water', pp. 330–1.
67. Ibid., pp. 331–2.
68. Ibid., p. 333.
69. Ibid., p. 334.
70. See Chapter 5.
71. Ibid., p. 335.
72. Ibid.
73. Ibid., p. 335.
74. Ibid., p. 336.
75. Watt, 'Sequel to the Thoughts on the Constituent Parts of Water'.
76. Ibid., p. 357.
77. The Common Place Book is a thick folio notebook, bound in vellum, with entries covering the period 1782–1812, Archives of Soho, MS3219/4/171.
78. J. Priestley, *Experiments and Observations on Different Kinds of Air...In Three Volumes*, 3rd edn (Birmingham: Thomas Pearson, 1790), vol 1, p. 282.
79. Once we are attuned to the rhetorical aspects of scientific practice, we might think that Priestley was here using Watt's intervention as a device to feign a more balanced and rational consideration of Lavoisier's ideas on his own part than was the case.
80. See Golinski, *Science in Public Culture*, pp. 130–7.
81. Watt to Black, 11 November 1784, in Robinson and McKie (eds), *Partners in Science*, p. p. 145–7. In this respect Watt presumably saw himself as engaged in the same sorts of conversations as the regulars at the meetings of the Coffee House Philosophical Society, even though he did not attend their meetings. See J. Golinski, 'Conversations on Chemistry: Talk about Phlogiston in the Coffee House Society, 1780–87', in T. H. Levere and G. L'E. Turner (eds), *Discussing Chemistry and Steam: The Minutes of a Coffee House Philosophical Society* (Oxford: Oxford University Press, 2002), pp. 191–205.
82. Watt to Black, 13 August 1786, in Robinson and McKie (eds), *Partners in Science*, pp. 152–3.
83. Watt to Black, 4 February 1788, in Robinson and McKie (eds), *Partners in Science*, pp. 158–9.
84. R. Kirwan, *An Essay on Phlogiston, and the Constitution of Acids. A New Edition* (London: J. Johnson, 1789), p. 43. Kirwan, however, disagreed with Watt's account of the production of dephlogisticated air from nitre (see pp. 94–5, 113). On Kirwan and his importance, see V. Boantza, 'The Phlogistic Role of Heat in the Chemical Revolution and the Origins of Kirwan's 'Ingenious Modifications ... Into the Theory of Phlogiston' ', *Annals of Science*, 65 (2008), pp. 309–38.
85. J. Priestley, 'Additional Experiments and Observations Relating to the Principle of Acidity, the Decomposition of Water, and Phlogiston...With Letters to Him on the Subject by Dr Withering and James Keir Esq. ', *Philosophical Transactions*, 81 (1791), pp. 213–22.
86. W. H. Brock, 'Joseph Priestley, Enlightened Experimentalist', in I. Rivers and D. L. Wykes (eds), *Joseph Priestley, Scientist, Philosopher, and Theologian* (Oxford: Oxford University Press, 2008), pp. 49–79.

87. Schofield (ed.), *A Scientific Autobiography*, pp. 198ff.
88. Hills, *James Watt, Volume 2*, p. 223.
89. Priestley to Watt, 29 April 1783, in J. P. Muirhead (ed.), *Correspondence of the Late James Watt on his Discovery of the Theory of the Composition of Water* (London: John Murray, 1846), pp. 25–6.
90. Watt to Priestley, 2 May 1783, in Muirhead (ed.), *Correspondence of the Late James Watt*, pp. 26–8 on p. 27.
91. Watt to Black, 3 February 1783, in Robinson and McKie (eds), *Partners in Science*, pp. 121–2 on p. 122. Richard Hills informs us of Watt's puzzlement in 1770 by the great amount of air in his condensing apparatus and that in his early engines 'he made ample provision for the removal of air and water with triple pumps'. (See Hills, *James Watt. Volume 2*, p. 220.)
92. Priestley, *Experiments and Observations on Different Kinds of Air...* vol 2, p. 409.
93. The numerous 'ofs' in that sentence are the source of confusion. One possible reading is that it is 'the water of Air' (that is water contained in air) that was extricated. But a little punctuation gives what I think must be the correct meaning: '...the 10th of the bulk, of the water, of Air extricated...'.
94. Muirhead, *Life of Watt*, p. 399.
95. The episode is described in detail in Musson and Robinson, *Science and Technology in the Industrial Revolution*, pp. 251–337. See also Hills, *James Watt. Volume 2*, pp. 137–47.
96. D. F. Larder, 'An Unpublished Chemical Essay of James Watt', *Notes and Records of the Royal Society of London*, 25 (1970), pp. 193–210.
97. Watt and Berthollet had an extensive correspondence. For part of their nomenclatural discussions see Watt to Berthollet, 18 [December] 1788, Archives of Soho, MS3219/4/123, 268), Watt to Berthollet, 24 January 1789 (MS3219/4/123, 279) and Berthollet to Watt, 28 December 1788 (MS3219/4/102).
98. Larder, 'An Unpublished Chemical Essay', p. 194.
99. Ibid., p. 202.
100. Ibid., p. 203.
101. See Beddoes and Watt, *Considerations*, and discussion in Miller, *Discovering Water*, pp. 53–4 and in Miller and Levere, 'Inhale it and See?', on which the following sections draw extensively.
102. The publication history of the *Considerations* was a complex one, as successive editions incorporated new parts and revised and extended old ones. The first edition was: *Considerations on the Medicinal Use of Factitious Airs, and the Manner of Obtaining them in Large Quantities. In two parts. Part I by Thomas Beddoes, M. D. Part II by James Watt, Esq.* (Bristol, 1794). A second edition was published as *Considerations on the Medicinal Use, and on the Production of Factitious Airs Part I by Thomas Beddoes, M. D. Part II by James Watt, Engineer. Edition the Second, to which are Added Communications from Doctors Carmichael, Darwin, Ewart ... and Others* (Bristol: Bulgin and Rosser, 1795). In 1795 a Part III was also published detailing more medical cases. In 1796, was published: *Considerations on the Medicinal Use, and on the Production of Factitious Airs. Part I by Thomas Beddoes M. D. Part II by James Watt, Engineer. Edition the Third, Corrected and Enlarged* (Bristol: J. Johnson and H. Murray, 1796). All reference and analysis below uses this third edition.
103. See J. E. Stock, *Memoirs of Thomas Beddoes, M. D. with an Analytical Account of his Writings* (1811; Thoemmes Press, 2003), pp. 104–5, 133–4; Muirhead, *Life of Watt*,p. 429; Stansfield and Stansfield, 'Dr Thomas Beddoes and James Watt'; F. F. Cartwright, 'The

Association of Thomas Beddoes, M. D., with James Watt, F. R. S.', *Notes and Records of the Royal Society of London*, 22 (1967), pp. 131–43.

104. A very important exception is Golinski, *Science as Public Culture*, pp. 153–87.
105. Beddoes to Black, 15 April 1791, Edinburgh University Library, Special Collections, MS Gen 873/III/200,201
106. See, for example, Beddoes to Black 6 November 1787, Edinburgh University Library, Special Collections, MS Gen 875/III/52,53; Joseph Black, lecture notes ca. 1785, Royal College of Physicians of Edinburgh, MS M9. 42, M9. 43.
107. Beddoes to Joseph Black, 6 November 1787, Edinburgh University Library, Special Collections, Black Correspondence, Gen 875/IIIF, ff. 52–3.
108. Beddoes to Black, 21 April [1789], Edinburgh University Library, Special Collections, Black Correspondence Gen 873/III/f. 129–130.
109. T. H. Levere, 'Dr Thomas Beddoes (1760–1808): Chemistry, Medicine, and Books in the French and Chemical Revolutions', in L. Principe (ed.), *New Narratives in Eighteenth-Century Chemistry*, (Dordrecht: Springer, 2007), pp. 157–76.
110. Watt to Joseph Black, 17 July 1793, in Robinson and McKie (eds), *Partners in Science*, pp. 195–6, on 196.
111. T. Beddoes, 'Reasons, For believing the friends of liberty in France not to be the authors or abettors of the crimes committed in that country', 9 October 1793, copy in Cornwall Record Office, DG 41/25CRO. See T. H. Levere, 'Dr. Thomas Beddoes at Oxford. Radical politics in 1788–1793 and the fate of the regius chair in chemistry', *Ambix*, 28 (1981), pp. 61–9, and idem, 'Dr. Thomas Beddoes (1760–1808): Science and Medicine in Politics and Society', *The British Journal for the History of Science*, 17 (1984), pp. 187–204.
112. Hills, *James Watt. Volume 3*, pp. 152–8.
113. Ibid., p. 153.
114. Several members of the Lunar Society were major contributors, the Wedgwoods more prominent than the Watts in this regard. See T. H. Levere, 'Dr. Thomas Beddoes and the Establishment of his Pneumatic Institution. A Tale of Three Presidents', *Notes and Records of the Royal Society of London*, 32 (1977), pp. 41–9. The most extraordinary funding, a bond for £10,000 pounds, came from James Keir in Bond for performance of Covenant. Dr Beddoes to Jas. Keir & Wm. Reynolds Esqrs., 27 August 1793, emended to 16 April 1794, National Archives (Kew), MSC 104/41. Keir was also a staunch phlogistonist.
115. Josiah Wedgwood II to Thomas Poole, British Library Add MSS. 35, 345 ff. 148–149, dated 1799/12/19 describes Beddoes's use of a warm room and oxygenating the air.
116. Watt to Beddoes, 6 May 1803, Archives of Soho, MS 3219/4/118, 538.
117. See Hills, *James Watt. Volume 3*, p. 159.
118. Beddoes adopted the concept of irritability from his friend Christoph Girtanner (1760–1800) who was a fellow student at Edinburgh in the mid-1780s. See N. Vickers, *Coleridge and the Doctors 1795–1806* (Oxford: Clarendon Press, 2004), pp. 55–9.
119. Between mid-1792 and late 1793 Beddoes published a number of works, which set out the theoretical basis for his plans to develop pneumatic medicine. That basis lay primarily in the notion of regulating the atmosphere, that is, adjusting (in experimental and clinical situations) the amount of oxygen that the breathed atmosphere contained, or at least the amount of oxygen taken in by patients. See, in particular, T. Beddoes, *Observations on the Nature and Cure of Calculus, Sea Scurvy, Consumption, Catarrh and Fever: Together with Conjectures upon Several other subjects of Physiology*

and Pathology (London: J. Murray, 1793) and T. Beddoes, *A Letter to Erasmus Darwin, M. D. on a New Method of Treating Pulmonary Consumption, and some Diseases hitherto found Incurable* (Bristol: Bulgin & Rosser, 1793), and also *Letters from Dr. Withering...Dr. Ewart...Dr Thornton... and Dr. Biggs... together with some other Papers, Supplementary to two Publications on Asthma, Consumption, Fever, and other Diseases by T. Beddoes* (Bristol: n. p., 1794).

120. See Beddoes and Watt, *Considerations*, Part III, 2nd edn, 1796, pp. 1–3.
121. Lavoisier's estimation of atmospheric oxygen, in his *Traité* (1789) and elsewhere was around 20% too high; Beddoes was accordingly working with an oxygen-enriched air.
122. Beddoes and Watt, *Considerations* (1796), pp. 107–18, 181–219. Hereafter the account of the apparatus (pp. 181–219) I will refer to as 'Apparatus', and the other contribution (pp. 107–18) I will refer to as 'Watt's hints'.
123. 'Watt's hints', p. 107.
124. 'Apparatus', p. 182.
125. 'Watt's hints', p. 108.
126. J . R. Partington, *History of Chemistry*, 4 vols (London: Macmillan, 1962–4), vol. 3, pp. 345–62; D. R. Dyck, 'The Nature of Heat'.
127. 'Apparatus', pp. 212–13.
128. 'Watt's hints', p. 108. Here, Watt writes of different species of air, which are more than modifications – he seems to be balancing on the edge of the two different categorizations of airs.
129. 'Watt's hints', p. 108 (my italics).
130. 'Apparatus', p. 212.
131. Whilst here Watt treats 'inflammable air, properly so called' and 'charcoal in a state of solution' as different, he elsewhere speculates that light inflammable air ('hydrogene') is in fact a solution of charcoal in water.
132. 'Watt's hints', p. 109.
133. But compare this view with that expressed in November 1794 (letter for Percival) in which Watt refers to the 'positive powers' of fixed air in this respect. Watt's account here explicitly equates solution with chemical combination.
134. 'Watt's hints', p. 109.
135. C-L. Berthollet, *Recherches sur les Lois de l'Affinité* (Paris: Badouin, 1801).
136. On the earlier origins of Berthollet's ideas on affinity see: Partington, *History of Chemistry*, vol. 3, p. 515; P. Grapí and M. Izquierdo, 'Berthollet's Conception of Chemical Change in Context', *Ambix*, 44 (1997), pp. 113–30, who locate them in the years 1789–1793 in the context of nitre production, and B. W. Keyser, 'Between Science and Craft: The Case of Berthollet and Dyeing', *Annals of Science*, 47 (1990), pp. 213–60, esp. 234–5 which agrees with Partington that Berthollet's *Eléments de l'Art de la Teinture*, 2 vols (Paris, 1791) was important in the development and application of those ideas. It appears that these ideas also became part of the currency of the debate between the phlogistians and anti-phlogistians. Priestley in *The Doctrine of Phlogiston Established* (1800) states: 'I once thought with the Antiphlogistians that fixed air was held in solution in the heavy inflammable air, because when it is fired together with dephlogisticated air much fixed air is produced; but I am now satisfied that all the oxygen in that fixed air comes from the dephlogisticated air with which it is fired' (p. 52).
137. Lavoisier, *Traité* (1789), vol. 1, p. xiv. For Berthollet, see below.
138. 'Letter intended to be written to Dr Percival Novr 1794 on the subject of airs'. The draft of this letter is dated Birmingham 24 November 1794, Archives of Soho, MS3219/4/27.

139. Watt to Beddoes, 4 March 1795, Archives of Soho, MS3219/4/124. This quotation is used (in a slightly different, inaccurate form) in Hills, *James Watt. Volume 3*, p. 156.
140. Watt to Beddoes, 25 October 1794, Archives of Soho, MS3219/4/124. Telling Erasmus Darwin of Gregory's departure, Watt said: 'Gregory being gone to College, I make no experiments being only half alive'. Watt's illness was caused in part by worry over patent trial business, which preoccupied him in the winter of 1794–5. Gregory had gone to Glasgow College. (Watt to Darwin, 11 November 1794, Archives of Soho, MS3219/4/124.)
141. Watt to Darwin, July? 1794, Archives of Soho, MS3219/4/124. In his previous letter to Darwin of 30 June, Watt had referred to the process of *expelling* 'inflammable air from red hot zinc by water', which signalled to Darwin that Watt might still be adhering to anti-Lavoisian ideas. Darwin, who had abandoned belief in phlogiston himself, inquired gently: '...you speak of obtaining inflammable air from Zinc. Do you hold the theory that the inflammable air (Hydrogen) is extracted from the Zinc, as your expression seems [to] countenance, or do you suppose that it is from the decomposition of water according to the idea of poor Lavoisier?' [Darwin to Watt, 3 July 1794, in King-Hele (ed.), *Collected Letters of Darwin*, p. 441]. It was in response to that inquiry that Watt declared his continuing allegiance to phlogiston.
142. Watt to Dr. Ewart, 21 September 1794, Archives of Soho, MS3219/4/124.
143. Watt to Beddoes, 30 August 1794, Archives of Soho, MS3219/4/124.
144. Watt to Beddoes, 4 September 1794, Archives of Soho, MS3219/4/124.
145. J. Priestley, *Experiments on the Generation of Air from Water* (London: J. Johnson, 1793), p. 12.
146. See Schofield, *The Enlightened Joseph Priestley*, pp. 353–68.
147. J. Priestley, *The Doctrine of Phlogiston Established, and that of the Composition of Water Refuted* (Northumberland, PA: Printed for the Author, 1800), pp. 51–2.
148. Other, better known, extenders of Black's work in this sense were, of course, William Irvine, Adair Crawford and William Cleghorn on whom see Donovan, *Philosophical Chemistry*, pp. 271ff. See also Fox, *The Caloric Theory of Gases*, Chapter 1.

5 The Steam Engine as Chemistry

1. In the case of Priestley, Schofield long ago suggested that awareness of such greater coherences showed that there was a 'connectedness' in Priestley's experimental interests that was and is unappreciated by his many critics. See Schofield, *A Scientific Autobiography*, p. 268.
2. W. C. Patterson, *Nuclear Power*, 2nd edn (Harmondsworth: Penguin, 1983), p. 78 recounts the accidental death in May 1946 of Louis Slotin, a technician at Los Alamos. Slotin was performing a manual procedure with a screwdriver to bring two hemispheres of Uranium-235 to the point of criticality. This he called 'twisting the tail of the dragon'. Unfortunately for Slotin his screwdriver slipped.
3. See L. Roberts, 'An Arcadian Apparatus. The Introduction of the Steam Engine into the Dutch Landscape', *Technology and Culture*, 45 (2004), pp. 251–76.
4. Watt to William Small, 17 August 1773, quoted in Schofield, *The Lunar Society*, p. 70, and in W. E. K. Middleton, *A History of the Theories of Rain and other Forms of Precipitation* (London: Oldbourne, 1965), p. 37.

5. Journal of James Watt, May 1773–March 1774, pp. 51–4, 107 records Watt's reading of De Luc, his attempt to construct a barometer like De Luc's and experiments on the boiling of water. (Archives of Soho, MS3219/4/135.)
6. Watt to Small, 17 August 1773, in Muirhead, *Origin and Progress*, vol. 2, pp. 58–9, on p. 59. Note that here Watt uses the term 'perfect engine' to mean the abstract engine, that which could be revealed without giving too much away. These matters had been discussed between Watt and Small since before Watt's initial patent application for the separate condenser in 1769.
7. Friedrich Wilhelm, Baron Reden (1752–1815), Hanoverian Statesman, was among the first of a long line of visitors admitted to the Soho manufactory whose presence often required nice judgements about the boundary between philosophical interest and industrial espionage. See P. M. Jones, 'Industrial Enlightenment in Practice. Visitors to the Soho Manufactory, 1765–1820', *Midland History*, 33 (2008), pp. 68–96.
8. Schofield, *The Lunar Society*, p. 240; Watt to Black, 13 December 1782, in Robinson and MacKie, *Partners in Science*, pp. 117–18.
9. Schofield, *The Lunar Society*, pp. 240–1.
10. Watt to Black, 13 December 1782, in Robinson and MacKie, *Partners in Science*, pp. 117–18.
11. Watt to Matthew Boulton, 10 December 1782, in Muirhead, *Origin and Progress*, vol 2, pp. 167–8.
12. Watt's Diary for 1783 records De Luc as arriving on 6 March and leaving on 16 March. (Archives of Soho, MS3219/4/142.) In a letter to Black after the fact of De Luc's visit Watt stated that 'Mr De Luc came according to his promise and staid 10 days'. (Watt to Black, 21 April 1783, in Robinson and MacKie (eds), *Partners in Science*, pp. 124–7.)
13. Watt to Black, 21 April 1783, in Robinson and McKie (eds), *Partners in Science*, on pp. 124–5.
14. Another experiment involving airs that was intended to give insight into meteorological phenomena, was made by Boulton and Watt in 1784. Though generally suspicious of the thirst for empty spectacle activating most 'balloonomania', as Watt called it, he did see potential value for natural philosophical inquiry. A balloon filled with a mixture of common and inflammable air was equipped with a lighted fuse and launched. The intention was 'to determine whether the Growling of thunder is owing to echos or to successive explosions'. Unfortunately, a loud shout from the crowd, which had assembled to see the experiment interfered with the experimenters' hearing of the explosion of the balloon and the immediate aftermath of it. (See Hills, *James Watt. Volume 2*, pp. 214–5.) The crowd's frustration of his philosophical purpose neatly manifested the conflict that Watt and others saw between balloons as spectacle and as instruments of philosophical inquiry.
15. De Luc, *Idées sur la Météorologie*, 2 vols (London: T. Spilsbury, 1786–7), vol. 1, pp. 215–16. On De Luc's reliance on Scottish pneumatic chemistry and the growing hegemony of chemistry in meteorology at this time see V. Jankovic, *Reading the Skies: A Cultural History of English Weather, 1650–1820* (Manchester: Manchester University Press, 2000), pp. 151–4. That reliance is treated in more detail in T. S. Feldmann, 'The History of Meteorology, 1750–1800: A Study in the Quantification of Experimental Physics', Unpublished PhD Dissertation, University of California, Berkeley, 1983.
16. De Luc, *Idées*, quoted in W. E. K. Middleton, 'Chemistry and Meteorology, 1700–1825', *Annals of Science*, 20 (1965), pp. 125–41 on pp. 130–1.
17. De Luc, *Idées*, para 672, as quoted in Middleton, 'Chemistry and Meteorology', p, 132.

18. De Luc, *Idées*, para 675, as quoted in Middleton, 'Chemistry and Meteorology', p. 132. As Middleton points out, De Luc presumably really meant to refer to 'land birds'!
19. See Chapter 4 above.
20. Feldmann, 'The History of Meteorology, 1750–1800', pp. 46–7.
21. Watt to Gilbert Hamilton, 28 September 1783, Archives of Soho, MS3219/4/123, 42. I am assuming that the publication to which Watt refers here was what eventually became De Luc's *Idées sur la Météorologie*.
22. It is not clear, for example, that in breaking the news of De Luc's theory to Hamilton Watt was also showing that the theory was new to him. But if it *was* new to him in 1783 he could hardly have been explicitly rejecting it in the letters of 1779 and 1780 that Feldmann uses to suggest that he did. Indeed those letters may simply have been referring to what Watt regarded as a standard view of things at the time. Of course, it is also true that Watt's desire to give De Luc every assistance in advancing his ideas did not necessarily mean that Watt endorsed them.
23. Quoted in Jankovic, *Reading the Skies*, p. 154. Jankovic points to a range of other authors who likened the atmosphere to a chemical laboratory at this time, including Le Roy, George Adams, Antoine François Fourcroy, Richard Kirwan, George Shaw, William Nicholson and Thomas Thomson.
24. Gerstner, 'James Hutton's Theory of the Earth'; Donovan, 'James Hutton', pp. 176–7.
25. See especially, Playfair, *Illustrations of the Huttonian Theory of the Earth* and Oldroyd, *Thinking about the Earth*, pp. 86–107.
26. Oldroyd, *Thinking about the Earth*, p. 93.
27. Anon, 'Matthew Boulton', in *Public Characters of 1800–1801*, pp. 1–9, p. 3. The reference to the leviathan draws on the Book of Job 42:1–2, concerning the relations of man and nature, which is I think the appropriate context here. The other context for such phrases is in relation to the taming of the Hobbesian State, usually by the law.
28. J. Jones, H. S. Torrens and E. Robinson, 'The Correspondence between James Hutton (1726–1797) and James Watt (1736–1819) with two letters from Hutton to George Clerk-Maxwell (1715–1784): Part I', *Annals of Science*, 51 (1994), pp. 637–53, on p. 638.
29. Ibid., p. 639. Jones et al. believe that Hutton and Watt visited Cheshire as a detour on their way south to Birmingham.
30. Robinson and McKie (eds), *Partners in Science*, pp. 173, 186–7, 197, 205, 208, 221.
31. Hutton's presence is mentioned in E. Darwin, 'Frigorific Experiments on the Mechanical Expansion of Air', *Philosophical Transactions of the Royal Society of London*, 78 (1788), pp. 43–52 on p. 44.
32. Watt wrote about these experiments to William Small in August 1773, just a few months before his journey with Hutton. This was the time when Watt also considered writing a book on 'Elements of the Theory of Steam-Engines'.
33. See the review of Playfair, *Illustrations of the Huttonian Theory* in *The Monthly Review*, series 2, 42 (1803), pp. 285–94 on pp. 287–8.
34. J. Playfair, 'Biographical Account of the late Dr James Hutton, F. R. S. Edin. ', *Transactions of the Royal Society of Edinburgh*, 5 (1805), pp. 39–99 on p. 91. Playfair's account is also notable for the extent to which it emphasises the centrality of chemistry to Hutton's intellectual world: '...he [Hutton] began from chemistry; and it was from thence that he took his departure in his circum-navigation both of the material and intellectual world' (p. 74). This is ironic given Playfair's role in disguising the details of Hutton's chemistry of heat and their role in his theory.

35. G. L. Davies, *The Earth in Decay: A History of British Geomorphology 1578–1878* (London: Macdonald Technical and Scientific, 1969), pp. 175–6. The theme of Watt's steam engine as inspiration for Hutton is pursued enthusiastically in S. Baxter, *Revolutions in the Earth. James Hutton and the True Age of the World* (London: Weidenfeld & Nicolson, 2003), pp. 98–108.
36. See D. R. Dean, 'Hutton Scholarship, 1992–1997', in G. Y. Craig and J. H. Hull (eds), *James Hutton — Present and Future* (London: Geological Society of London Special Publications, 1999), pp. 175–179 on p. 176. Dean's targets in his denial of the accuracy of this attribution to Hutton are D. Allchin, 'James Hutton and Phlogiston', *Annals of Science*, 51 (1994), pp. 615–35 and Oldroyd, *Thinking about the Earth*.
37. Davies, *Earth in Decay*, p. 176.
38. J. Hutton, *Theory of the Earth, with Proofs and Illustrations* 2 vols (Edinburgh: William Creech, 1795), vol. 1, p. 146. In this passage Hutton is denigrating the common practice of finding in geological activity warnings or punishments from God. A focus of debates between religious and naturalistic explanations of earthquakes had long been the Lisbon earthquake of mid- century. See J. Kosák, V. Moreira and D. Oldroyd, *The Iconography of the Lisbon Earthquake* (Prague: Geophysical Institute of the Academy of Sciences of the Czech Republic, 2005).
39. J. Hutton, 'Theory of the Earth; or an Investigation of the Laws observable in the Composition, Dissolution, and Restoration of Land upon the Globe', *Transactions of the Royal Society of Edinburgh*, vol 1, Pt II, (1788), pp. 209–304, on pp. 215–6.
40. Schofield (ed.), *A Scientific Autobiography*, p. 267.
41. See Priestley's covering letter to W. Withering, 'An Analysis of Two Mineral Substances, viz. the Rowley-rag-stone and the Toad-stone', *Philosophical Transactions of the Royal Society of London*, 72 (1782), pp. 327–36, on p. 327; Priestley, *Experiments and Observations Relating to Natural Philosophy ... III*, pp. 216–23.
42. J. Priestley, 'Observations on Different Kinds of Air', *Philosophical Transactions of the Royal Society of London*, 62 (1772), pp. 147–264 on p. 162, quoted in Schofield (ed.), *A Scientific Autobiography*, p. 268.
43. Boulton to Watt, 21 July 1781, Archives of Soho, MS3147/3/5. The meanings of the symbols are: >⊖ marine acid; ⨸^ volatile alkali; ▽ water.
44. Ibid.
45. 'Simcox' is difficult to identify. Perhaps he was the W. J. Simcox mentioned in a letter from Henderson to Boulton, 17 August 1777, who was engaged in engine innovations. Or, more likely, he is the Simcock who proposed using 'an elastic vapour instead of steam', who is mentioned as a projector in Dickinson and Jenkins, *James Watt and the Steam Engine*, pp. 322–3.
46. Ibid.
47. Archives of Soho, MS 3219/4/287/7 (folder 2).
48. Both Watt and De Luc were associated with Robert Jameson's Wernerian Society and so apparently allied with the Neptunist faction. But we know that Watt, at least, was a long-time friend and admirer of James Hutton, the figurehead of the Vulcanists. The dichotomy in any case is a false one since all agreed that both fire and water had their role.
49. J. Robison, 'Steam', p. 735. Robison drew an elaborate analogy between the 'succussion' that occurs as bubbles collapse at the bottom of a boiling kettle and the phenomena of earthquakes. He does however note that there are objections to the idea.

50. Erasmus Darwin to Matthew Boulton, 12 December [17]65, Archives of Soho, MS 3782/13/53/30, as transcribed in King-Hele (ed.), *Collected Letters of Darwin*, pp. 65–6. I have omitted Dr. King-Hele's interpolation of 'at' after 'as' in his transcription of this letter. This interpolation is an understandable but I think erroneous 'correction' or unnecessary amplification. This is so because Darwin's concern is with the co-variation of the size of the surface at which the steam is produced and the quantity of that steam. So throughout the letter he refers to quantity varying either 'as' the upper surface of the boiling water or 'as' the surface of the vessel in contact with the fire. The location of the production of the steam ('at. . ') is implicit in the covariation ('as...').
51. Ibid., p. 66. In a jocular, but nevertheless telling, analogy in this same letter Darwin compared the way the strength and activity of man depended on the 'Quantity of vital Steam rising into the Brain from his boiling Blood' with the way that 'this Animal' [the steam-engine] depended on the design of the boiler.
52. Watt's notebook in 1765 recorded studies of the boiler of Jonathan Sisson's engine: '...I saw no boyler so perfect... as the common tea kitchen [kettle].... Here the fuel is always in contact with the sides of the boyler containing the water....'. This is consistent with the view that the significant factor in producing steam is to ensure the maximum proximity of heat (fire particles) and water. The Notebook is reproduced in Robinson and McKie (eds), *Partners in Science,* the section in question being on p. 435.
53. Darwin, 'Frigorific Experiments'.
54. Ibid., p. 50.
55. Ibid., pp. 50–1.
56. See Chapter 2, above.
57. E. Darwin, *The Botanic Garden. A Poem in Two Parts*. Pt 1, Canto 1, lines 253–62. It is tempting to read into the last line – '...and shakes the earth' – an allusion to the geological functions of steam, which are discussed in other parts of Darwin's poem.
58. See Chapter 2 above.
59. Darwin, *The Botanic Garden*, note VII 'Elementary Heat'.
60. J. Riskin, *Science in the Age of Sensibility: The Sentimental Empiricists of the French Enlightenment* (Chicago, IL: University of Chicago Press, 2002). Although Riskin's account of 'sentimental empiricism' is based in the French context, she does seem open to it being manifest elsewhere, in fact wherever the conflict over the new rationalist language of Lavoisian chemistry and the language of phlogiston was alive. Riskin mentions Hutton in this regard (p. 248).
61. Recall how Watt, in his comments on the French nomenclature, was willing to retain the term 'caloric' but gave it a new meaning consonant with his ideas about the chemistry of heat. See above, Chapter 4.
62. The Notebook is published in Robinson and McKie (eds), *Partners in Science*, pp. 431–90. The original is in the Archives of Soho, MS 3219/4/170.
63. Robinson and McKie (eds), *Partners in Science*, p. 439.
64. On the workings of the Newcomen Engine see R. L. Hills, *Power from Steam: A History of the Stationary Steam Engine* (1989; Cambridge: Cambridge University Press, 1993), pp. 20–30.
65. Watt, 'Plain Story', in Muirhead, *Life of Watt*, p. 87.
66. R. L. Hills, 'The Origins of James Watt's Perfect Engine', *Transactions of the Newcomen Society*, 68 (1996–7), pp. 85–107; idem, 'How James Watt Invented the Separate Condenser', *Bulletin of the Scientific Instrument Society*, 57 (1998), pp. 26–9 and 58 (1998),

pp. 6–10; idem, *James Watt. Volume 1*, pp. 363–86. Another very clear exposition of Hills' approach is in B. Marsden, *Watt's Perfect Engine: Steam and the Age of Invention* (Cambridge: Icon Books, 2002), pp. 43–64.

67. Hills, *James Watt. Volume 2*, p. 27.
68. Marsden, *Watt's Perfect Engine*, p. 51.
69. Robinson and McKie, (eds), *Partners in Science*, p. 459.
70. This is the Table devised by Watt in 1814 as rendered in Robinson and McKie (eds), *Partners in Science*, p. 473. Watt had made some corrections to the results as originally recorded in the Notebook but these do not materially affect our concerns here.
71. D. Fleming, 'Latent Heat and the Invention of the Watt Engine', *Isis*, 43, (1952), pp. 3–5.
72. This treatment of these issues relies heavily on, whilst differing in emphasis from, Donovan, *Philosophical Chemistry*, pp. 250–65.

6 The Indicator Understood, or Why Watt was not a Proto-thermodynamicist

1. K. J. Laidler, 'James Watt and the Science of Thermodynamics', in idem, *To Light Such a Candle. Chapters in the History of Science and Technology* (Oxford: Oxford University Press, 1998), pp. 13–67. See also the more sophisticated treatment by D. V. Fenby, 'Chemical Reactivity and Heat in the Eighteenth Century', in P. Jones (ed.), *Philosophy and Science in the Scottish Enlightenment* (Edinburgh: John Donald Publishers, 1988), pp. 67–86. Fenby concludes that 'the influence of the Scottish Enlightenment and, in particular, of Watt and Black on the development of thermodynamics was both direct and considerable' (p. 83). The connections that he finds, so far as they have substance, were, I suggest, created by nineteenth century *users*, rather than being 'influences'.
2. See Cardwell, *From Watt to Clausius*, pp. 42, 54.
3. Hills, 'How James Watt Invented the Separate Condenser'; idem, 'The Origins of James Watt's Perfect Engine'. See also idem, *James Watt. Volume 1*, pp. 363–86.
4. Even this becomes doubtful if it is true that Watt's own accounts, such as the 'Plain Story', were activated as much by the objective of creating particular impressions in contested situations as they were by giving a 'true' history. I have argued in my account of Watt's 'self-fashioning' (Chapter 2 above) that they were so activated.
5. For these, and other, local commemorations of Watt see R. M. Smith, *The History of Greenock* (Greenock: Orr, Pollock & Co., 1921), pp. 331–40.
6. The list of Watt Anniversary lecturers after Joule included many notables: John Scott Russell (twice), James Clerk Maxwell, Sir William Thomson (twice), Peter Guthrie Tait (twice), James Thomson (twice), Fleeming Jenkin, Balfour Stewart, Henry Dyer, Sir Frederick Bramwell, W. H. Preece, T. E. Thorpe, Sir Oliver Lodge, Sir William Ramsay, Frederick Soddy and Charles A. Parsons. The full list to 1911 is in *Greenock Philosophical Society, Jubilee Celebrations* (Greenock: The Philosophical Society, 1911). On the engineering, shipbuilding and academic context see MacLeod, *Heroes of Invention*, p. 347 and C. Smith, ' "Nowhere but in a Great Town": William Thomson's Spiral of Classroom Credibility', in C. Smith and J. Agar (eds), *Making Space for Science: Territorial Themes in the Shaping of Knowledge* (Basingstoke: Macmillan, 1998), pp. 118–46.

7. See D. S. L. Cardwell, *James Joule. A Biography* (Manchester: Manchester University Press, 1989), p. 183. How widely the Address was circulated, and therefore the extent to which its elisions 'influenced' others is difficult to gauge. It does seem that Joule at least circulated it among others of his thermodynamic circle. See, for example, Joule to William Thomson, March 1 [18]65, Kelvin Papers, Special Collections, University of Glasgow Library, J177.
8. J. P. Joule,'On Some Facts in the Science of Heat developed since the Time of Watt... Being a Lecture delivered before the Greenock Philosophical Society on 19th January, 1865', Papers of the Greenock Philosophical Society, No. 1, printed in *Publications of the Greenock Philosophical Society* (Greenock, 1887).
9. Ibid., p. 2.
10. Ibid., p. 15.
11. H. Dyer, 'The Steam Engine since the Days of Watt', *Papers of the Greenock Philosophical Society. –No. 10* (Greenock: William Hutchison, 1889). Dyer (1848–1918) was from a labouring family and after an engineering apprenticeship was a Whitworth Scholar at University of Glasgow. In 1873 he was appointed, on Rankine's recommendation, as Principal of the Imperial College of Engineering in Tokyo. After his return to Glasgow in 1882, Dyer became a prolific writer on, and advocate of, engineering education. At the time of his Watt Anniversary Address Dyer was Vice-President of the Institution of Engineers and Shipbuilders, Scotland. See O. Checkland, 'Dyer, Henry (1848–1918)', *Oxford Dictionary of National Biography*, online edition, Oxford University Press, 2004 [http://www.oxforddnb.com/view/article/40808].
12. Ibid., p. 10
13. Ibid., p. 12. The quotation continues: '...while our knowledge of the laws themselves have been developed in great part from the study of actual engines, and used to explain difficulties which had been overcome, and improvements which had been made from the results of practical experience'.
14. Both quotations are from Forbes, *Review of the Progress*, p. 71.
15. J. A. Ewing, 'Watt, James', *Encyclopaedia Britannica*, 9th edn, vol. 24 (1888), pp. 412–14, on p. 414.
16. The confidence that historians often have in the source of this statement is misplaced. The statement is attributed most often to Lawrence Joseph Henderson, the Harvard physiologist, historian and sociologist. J. B. Conant, *Science and Common Sense* (London: Oxford University Press, 1951), p. 39 tells us that Henderson was 'fond of remarking that before 1850 the steam engine did more for science than science did for the steam engine'. Conant was possibly the source for the quote in C. C. Gillispie, *The Edge of Objectivity* (Princeton, NJ: Princeton University Press, 1960), p. 357, where it is treated sceptically, and the admission of 'some truth' in it by R. J. Forbes, 'Power to 1850', in C. Singer et al. (eds), *A History of Technology* (Oxford: Oxford University Press, 1958), p. 165. Web-based sources frequently date the quotation to 1917 but it does not appear in Henderson's *The Order of Nature*, published in that year. None of the attributions to Henderson that I have seen refer to an alternative printed source. J. D. Bernal, *Science and Industry in the Nineteenth Century* (London: Routledge, 1953), p. 27, expresses similar sentiments. Lord Kelvin is sometimes attributed with a variant substituting 'thermodynamics' for 'science', but the source of this I also find remarkably fugitive, and I have seen it credited to J. Willard Gibbs.
17. Smith, *The Science of Energy*, p. 34.

18. See R. L. Hills and A. J. Pacey, 'The Measurement of Power in Early Steam-driven Textile Mills', *Technology and Culture*, 13 (1972), pp. 25–43.
19. Ibid., pp. 40–1.
20. Cardwell, *From Watt to* Clausius, pp. 80–3, 220–1; H. H. Jun., 'Account of a Steam-Engine Indicator', *Quarterly Journal of Science*, 13 (1822), pp. 91–6.
21. For Watt's description of the indicator see Robison, *System of Mechanical Philosophy*, vol. 2, pp. 156–7. On Farey and the indicator: R. B. Prosser, *Birmingham Inventors and Inventions* (Birmingham: The 'Journal' Printing Works, New Street, 1881), p. 36; John Farey in *Report of the Select Committee on the Law Relative to Patents for Inventions*, 1829, p. 138; A. P. Woolrich, 'John Farey and his *Treatise on the Steam Engine* (1827)', *History of Technology*, 22 (2000), pp. 63–106.
22. R. L. Hills, 'Houldsworth, Henry', in L. Day and I. McNeil (eds), *Biographical Dictionary of the History of Technology* (London: Routledge, 1996), pp. 356–7; T. Z. Penn, 'The Introduction of Calico Cylinder Printing in America: A Case Study in the Transmission of Technology', in I. M. G. Quimby and P. A. Earl (eds), *Technological Innovation and the Decorative Arts* (Charlottesville, VA: University of Virginia Press, 1974), pp. 247–50; D. J. Jeremy, *Artisans, Entrepreneurs, and Machines: Essays on the Early Anglo-American Textile Industries, 1770–1840s* (Aldershot: Ashgate, 1998).
23. Anon., 'Heartless Conduct of Henry Houldsworth Esq. ', *The Reformers' Gazette*, No. 143 (30 May, 1835), pp. 356–60; Anon., 'Another Slap at Henry Houldsworth', *The Reformers' Gazette,* No 145, (27 June 1835), 390–4. The case concerned a James Marsh whose inventions and improvements in machinery had allegedly been taken by Houldsworth without compensation.
24. Dickinson and Jenkins, *James Watt and the Steam Engine*, p. 230; Hills, *Power from Steam*, pp. 92–3.
25. Wedgwood's request was transmitted via William Murdock to Boulton & Watt, 5 February 1806. See Dickinson and Jenkins, *James Watt and the Steam Engine,* p. 231.
26. See J. Field, 'Joshua Field's Diary of a tour in 1821 through the Midlands', Intr. J. W. Hall, *Transactions of the Newcomen Society*, 6 (1925–6), pp. 1–41; also J. Field, 'Joshua Field's Diary of a tour in 1821 through the Provinces', Intr. E . C. Smith, *Transactions of the Newcomen Society*, 13 (1932–3), pp. 15–50.
27. É. Clapeyron, 'Mémoire sur la Puissance Motrice de la Chaleur', *Journal de l'Ecole Polytechnique*, cahier 23, 14 (1834), pp. 153–90. An edited translation of this paper is available in E. Mendoza (ed.), *Reflections on the Motive Power of Fire by Sadi Carnot and other Papers on the Second Law of Thermodynamics by E. Clapeyron and R. Clausius* (New York: Dover Publications, 1960), pp. 71–105.
28. Cardwell, *From Watt to Clausius*, pp. 220–1 is convinced that Clapeyron had encountered the indicator diagram but is not sure how. John Farey, as previously noted, had seen an indicator diagram being taken in Russia in 1819. It is possible that the same thing happened to Clapeyron on whose presence and activities in Russia at this time see M. Bradley, 'Franco-Russian Engineering Links: The Careers of Lamé and Clapeyron, 1820–1830', *Annals of Science*, 38 (1981), pp. 291–312.
29. N. L. S. Carnot, *Réflexions sur la puissance motrice du feu et sur les machines propres a developer cette puissance* (Paris: Bachelier, 1824).
30. See Smith and Wise, *Energy and Empire*, pp. 53, 290–1, 357–8.
31. P. G. Tait, *Heat* (London: Macmillan, 1884), pp. 298–301.
32. Sir James Alfred Ewing studied engineering at Edinburgh University under Fleeming Jenkin and P. G. Tait, and worked for William Thomson in making and laying submarine

cables. Like Henry Dyer, he taught at the Imperial University of Tokyo. After a spell as professor of engineering at University College, Dundee from 1883, in 1890 Ewing became professor of mechanism and applied mechanics at Cambridge University. See E. I. Carlyle, 'Ewing, Sir (James) Alfred (1855–1935), rev. W. H. Brock, *Oxford Dictionary of National Biography*, Oxford University Press, 204 [http://www.oxforddnb.com/view/article/33058].

33. Ewing, 'Watt, James', pp. 413–4.
34. See above Chapter 3, note 38. Marsden, 'Blowing Hot and Cold', pp. 390–400, offers important observations on the realities and appearances of the relations of theory and practice for Rankine and his associates.
35. W. H. Preece, *Watt and the Measurement of Power. Being the Watt Anniversary Lecture delivered before the Greenock Philosophical Society, 5 February 1897* (London: William Clowes & Sons, 1897). Preece (1834–1913) was Engineer-in-Chief of the Post Office, and an important, if controversial, figure in the electrical world of the late nineteenth century. See B. J. Hunt, 'Preece, Sir William Henry (1834–1913)', *Oxford Dictionary of National Biography*, Oxford University Press, 2004 [http://www.oxforddnb.com/view/article/35605]. See also E. C. Baker, *Sir William Preece, F. R. S., Victorian Engineer Extraordinary* (London: Hutchinson, 1976). It might be unfair to say that Preece's professed concern about attetntion to detail in the matter of instruments did not extend to his historical work. See G. J. N. Gooday, *The Morals of Measurement. Accuracy, Irony and Trust in late Victorian Electrical Practice* (Cambridge: Cambridge University Press, 2004), p. 68.
36. Ibid., p. 4.
37. See *The Times*, 30 March 1895, p. 19 under 'Sales by Auction', and *The Times*, 23 May 1895, p. 16 for the announcement of the sale in one lot.
38. I am very grateful to Dr Jim Andrew for showing me the Thinktank indicator during a visit there in November 2007, for subsequently photographing it, for further inspecting it to find the engraving, and also documenting its provenance. At the Cornwall Works of the Tangye family in Birmingham there was an exhibition devoted to James Watt in which the indicator was centrally displayed in a glass case. This is evident from a photograph of the exhibition on the 'Digital Handsworth' website (http://www.digitalhandsworth.org.uk/). The exhibition is dated to 1900– *c.* 1920, but was probably before 1911 since the artefact passed to Birmingham City Council along with the Boulton & Watt papers at that time. The Local Studies and History Department of Birmingham Central Library holds the original photograph.
39. This misdating was in fact an error on Preece's part. The question of who inscribed the misleading date on the indicator, when they did it, and why, remains open. One possibility is that this was done at the time of the sale of the instrument in 1895 and that the inscription reflected folklore internal to the Soho Foundry in the later nineteenth century identifying the instrument as Watt's 'original'.
40. D. Baird, 'Instruments on the Cusp of Science and Technology: The Indicator Diagram', *Knowledge and Society*, 8 (1989), pp. 107–22; D. Baird, *Thing Knowledge: A Philosophy of Scientific Instruments* (Berkeley, CA: University of California Press, 2004), pp. 41–3, 170–88.
41. Baird, 'Instruments', p. 121. Baird is, of course, drawing and building on recent emphasis by a number of scholars, including notably Peter Galison and Andy Pickering, upon the importance of 'practices' in science by way of critique of theory-centrism in accounts of science dynamics. See P. Galison, *Image and Logic: A Material Culture of Microphysics*

(Chicago, IL: University of Chicago Press, 1997); A Pickering, *The Mangle of Practice: Time, Agency and Science* (Chicago, IL: University of Chicago Press, 1995). However, given that instruments cannot speak for themselves, it seems to me that the idea of 'hardware-discourse couples' is a better concept than 'Thing Knowledge' in pursuing such questions. See, for example, J. A. Schuster and G. Watchirs, 'Natural Philosophy, Experiment and Discourse: Beyond the Kuhn/Bachelard Problematic', in H. LeGrand (ed.), *Experimental Inquiries* (Dordrecht: Kluwer, 1990), pp. 1–48.

42. Robison, 'Steam', p. 735.
43. Robison, *System of Mechanical Philosophy*, vol. 2, p. 14.
44. H. Cavendish and C. Blagden, 'Computations and Observations in Journey 1785', Devonshire Collections, Chatsworth, Cavendish Mss X(a) 4, pp. 36–7.
45. Dickinson and Jenkins, *James Watt and the Steam Engine*, pp. 228–9. Dickinson and Jenkins' account of the early indicators (pp. 228–33) has yet to be surpassed. But see also, J. Walter, *The Engine Indicator,* Chapter 1. In later chapters of this work Walter gives a valuable, comprehensive and well-illustrated account of subsequent developments in indicator design and practice.
46. Dickinson and Jenkins, *James Watt and the Steam Engine*, p. 228.
47. J. Robison, 'Steam-Engine', *Encyclopaedia Britannica*, 3rd edn (1797), vol. 17, pp. 758–9.
48. Hills, *James Watt. Volume 3*, p. 87.
49. A. J. Pacey, 'Some Early Heat Engine Concepts and the Conservation of Heat', *The British Journal for the History of Science*, 7 (1974), pp. 135–45, on p. 135.
50. Pacey, 'Some Early Heat Engine Concepts', p. 138.
51. Ibid., p. 142.
52. Ibid., p. 143.
53. Ibid., p. 143, quoting from P. Ewart, 'On the Measure of Moving Force', *Memoirs of the Manchester Literary and Philosophical Society*, 2nd series, 2 (1813), pp. 105–258. This is certainly an important paper so far as conceptual shifts are concerned, and there are signs that the Watts, father and son, reacted positively to it. (See D. S. L. Cardwell, 'Some Factors in the Early Development of the Concepts of Power, Work and Energy', *The British Journal for the History of Science* 3 (1967), pp. 209–224, p. 221.)
54. Ibid., p. 144.
55. Ibid.
56. Lee to James Lawson, 5 March 1798, quoted in Pacey, 'Some Early Heat Engine Concepts', p. 144.
57. R. Brain and M. N. Wise, 'Muscles and Engines: Indicator Diagrams and Helmholtz's Graphical Methods', in L. Krüger (ed.), *Universalgenie Helmholtz: Rückblick nach 100 Jahren* (Berlin: Akademie Verlag 1994), pp. 124–45; M. N. Wise, 'Work and Waste: Political Economy and Natural Philosophy in Nineteenth-Century Britain' (unpublished essay).
58. M. N. Wise, 'Precision: Agent of Unity and Product of Agreement. Part II – The Age of Steam and Telegraphy', in M. N. Wise (ed.), *The Values of Precision* (Princeton, NJ: Princeton University Press, 1995), pp. 222–36, on p. 233. Referred to is F. L. Holmes and K. M. Olesko, 'The Image of Precision: Helmholtz and the Graphical Method in Physiology', in Wise (ed.), *The Values of Precision*, pp. 198–221.

7 Conclusions

1. P. J. Bowler and I. R. Morus, *Making Modern Science. A Historical Survey* (Chicago, IL: University of Chicago Press, 2005), p. 402.
2. Rankine, *Introductory Lecture*, p. 22. See Chapter 3 above.
3. Jones, *Industrial Enlightenment*, pp. 16–17.

WORKS CITED

Manuscript Sources

Birmingham Central Library, Birmingham City Archives

Archives of Soho (James Watt Papers; Matthew Boulton Papers; Boulton & Watt Archives)

Cambridge University Library, Manuscripts and Archives

Kelvin Papers

Royal Greenwich Observatory, Airy Papers

Library of Congress, Washington D. C.

Andrew Carnegie Papers

National Archives (UK)

British Railways Board Files (AN)

Royal College of Physicians, Edinburgh

Royal Society, London

Certificates of Election and Candidature

Trinity College Library, Cambridge

Whewell Papers

University College London Library, Rare Books and Manuscripts

Brougham Papers

University of Edinburgh, Special Collections

Black Correspondence

University of Glasgow, Special Collections

Kelvin Papers

Muirhead Papers

University of St. Andrews Library

Forbes Papers

Electronic Resources

Oxford Dictionary of National Biography, http://www.oxforddnb.com

The Scotsman Digital Archive, http://edu.archive.scotsman.com/

The Times of London Digital Archive, http://www.gale.cengage.com/DigitalCollections/products/Times/

Printed Sources

Allchin, D., 'James Hutton and Phlogiston', *Annals of Science*, 51 (1994), pp. 615–35.

Anderson, R. G. W., *The Playfair Collection and the Teaching of Chemistry at the University ofEdinburgh 1713–1858* (Edinburgh: The Royal Scottish Museum, 1978).

—, 'Joseph Black and his Chemical Furnace', in idem, J. A. Bennett and W. F. Ryan, *Making Instruments Count: Essays on Historical Scientific Instruments Presented to Gerard L'Estrange Turner* (Aldershot: Ashgate Variorum, 1993).

—, 'Boerhaave to Black: The Evolution of Chemistry Teaching', *Ambix*, 53 (2006), pp. 237–54.

Anderson, W., *The Scottish Nation*, 9 vols (Edinburgh: Fullarton, 1875–1882).

Anon., 'Matthew Boulton', in *Public Characters of 1800–1801* (London: Richard Phillips, 1801), pp. 1–9.

—, 'Mr. James Watt', in *Public Characters of 1802–1803* (London: Richard Phillips, 1803), pp. 526–43.

—, 'Article IX. Analyses of Books', *Annals of Philosophy*, 1 (January–June 1813), pp. 459–65.

—, 'Recent Acquisitions in Edinburgh Museums, 1995–97: Supplement', *The Burlington Magazine*, 139 (1997), pp. 580–4.

Arago, F., *Historical Eloge of James Watt... Translated from the French... by James Patrick Muirhead* (London: John Murray, 1839).

Axon, W. E. A. (ed.), *The Annals of Manchester: A Chronological Record from the Earliest Times to the End of 1885* (Manchester: J. Heywood, 1886).

Baird, D., 'Instruments on the Cusp of Science and Technology: The Indicator Diagram', *Knowledge and Society*, 8 (1989), pp. 107–22.

—, *Thing Knowledge: A Philosophy of Scientific Instruments* (Berkeley, CA: University of California Press, 2004).

Baker, E. C., *Sir William Preece, F. R. S., Victorian Engineer Extraordinary* (London: Hutchinson, 1976).

Baldry, A. L., *Marcus Stone, R. A. His Life and Work* (London: Art Journal, 1896).

Barr, A., *James Watt and the Application of Science to the Mechanical Arts: An Inaugural Address Delivered in the University of Glasgow. November 11th, 1889* (Glasgow: James Maclehose & Sons, 1889).

Baxter, S., *Revolutions in the Earth. James Hutton and the True Age of the World* (London: Weidenfeld & Nicolson, 2003).

Beddoes, T., *A Letter to Erasmus Darwin, M. D. on a New Method of Treating Pulmonary Consumption, and some Diseases hitherto found Incurable* (Bristol: Bulgin & Rosser, 1793).

—, *Observations on the Nature and Cure of Calculus, Sea Scurvy, Consumption, Catarrh and Fever: Together with Conjectures upon Several other subjects of Physiology and Pathology* (London: J. Murray, 1793).

—, *Letters from Dr. Withering...Dr. Ewart...Dr Thornton... and Dr. Biggs... together with some other Papers, Supplementary to two Publications on Asthma, Consumption, Fever, and other Diseases by T. Beddoes* (Bristol: n. p., 1794).

—, and J. Watt, *Considerations on the Medicinal Use and the Production of Factitious Airs* (Bristol: J. Johnson and H. Murray, 3rd edn, enlarged, 1796).

Bernal, J. D., *Science and Industry in the Nineteenth Century* (London: Routledge, 1953).

Black, J., *Lectures on the Elements of Chemistry delivered in the University of Edinburgh by the late Joseph Black, M. D. ... Now Published from his Manuscripts by John Robison, LLD*, 2 vols (Edinburgh: William Creech, 1803).

Boantza, V., 'The Phlogistic Role of Heat in the Chemical Revolution and the Origins of Kirwan's 'Ingenious Modifications ... Into the Theory of Phlogiston'', *Annals of Science*, 65 (2008), pp. 309–38.

Bourne, J., *Handbook of the Steam-Engine: Containing all the Rules Required for the Right Construction and Management of Engines of Every Class* (London: Longman & Co., 1865).

Bowler, P. J. and I. R. Morus, *Making Modern Science. A Historical Survey* (Chicago, IL: University of Chicago Press, 2005).

Bradley, M., 'Franco-Russian Engineering Links: The Careers of Lamé and Clapeyron, 1820–1830', *Annals of Science*, 38 (1981), pp. 291–312.

Brain, R. and M. N. Wise, 'Muscles and Engines: Indicator Diagrams and Helmholtz's Graphical Methods', in Lorenz Krüger (ed.), *Universalgenie Helmholtz: Rückblick nach 100 Jahren* (Berlin, Akademie-Verlag, 1994), pp. 124–45.

Bramwell, F., 'Address by Sir Frederick Bramwell, D. C. L., F. R. S., M. INST. C. E., President', *Report of the Fifty-Eighth Meeting of the British Association for the Advancement of Science, held at Bath in September 1888* (London: John Murray, 1889).

Brock, W. H., 'The Society for the Perpetuation of Gmelin: The Cavendish Society, 1846–1872', *Annals of Science*, 35 (1978), pp. 599–617.

—, 'The Cavendish Society's Wonderful Repertory of Chemistry', *Annals of Science*, 47 (1990), pp. 77–80.

—, 'Joseph Priestley, Enlightened Experimentalist', in I. Rivers and D. L. Wykes (eds), *Joseph Priestley, Scientist, Philosopher, and Theologian* (Oxford: Oxford University Press, 2008), pp. 49–79.

Brougham, H., '[Review of] "Lectures on the Elements of Chemistry"', *Edinburgh Review*, 2 (1803), pp. 1–26.

Bryden, D. J., 'James Watt Merchant: The Glasgow Years, 1754–1774', in D. Smith (ed.), *Perceptions of Great Engineers: Fact and Fantasy* (London: Science Museum for the Newcomen Society, National Museums and Galleries on Merseyside and University of Liverpool, 1974), pp. 9–21.

Buchanan, R. A., *The Engineers: The History of the Engineering Profession in Britain* (London: Jessica Kingsley Publishers, 1989).

Caird, R., 'James Watt's Contribution to the Advancement of Engineering', Papers of the Greenock Philosophical Society (Greenock: 'Telegraph' Printing Works, [1910]).

Cantor, G. N., 'Henry Brougham and the Scottish Methodological Tradition', *Studies in the History and Philosophy of Science*, 2 (1971), pp. 69–89.

Cardwell, D. S. L., 'Some Factors in the Early Development of the Concepts of Power, Work and Energy', *The British Journal for the History of Science* 3 (1967), pp. 209–224.

—, *From Watt to Clausius. The Rise of Thermodynamics in the Early Industrial Age* (Ithaca, NY: Cornell University Press, 1971).

—, *James Joule. A Biography* (Manchester: Manchester University Press, 1989).

Carnegie, A., *James Watt* (Edinburgh and London: Oliphant, Anderson & Ferrier, 1905).

Carnot, N. L. S., *Réflexions sur la puissance motrice du feu et sur les machines propres a developer cette puissance* (Paris: Bachelier, 1824).

Carrington, N. T., *The King of the Golden River AND Letting off Steam. Two Plays for Juniors* (London: E. Mathews & Marott, 1936).

Cartwright, F. F., 'The Association of Thomas Beddoes, M. D., with James Watt, F. R. S.', *Notes and Records of the Royal Society of London*, 22 (1967), pp. 131–43.

Casteras, S. P., 'Excluding Women: The Cult of the Male Genius in Victorian Painting', in L. M. Shires (ed.), *Rewriting the Victorians: Theory, History, and the Politics of Gender* (London: Routledge, 1992), pp. 116–46.

Chalmers, G., *An Estimate of the Comparative Strength of Great-Britain, During the Present and Four Preceding Reigns...A new Edition; to which is prefixed, A Dedication to Dr James Currie, the Reputed Author of "Jasper Wilson's Letter"* (London: John Stockdale, 1794).

Champion, N., *James Watt* (Oxford: Heinemann, 2001).

Chang, H., *Inventing Temperature: Measurement and Scientific Progress* (Oxford: Oxford University Press, 2004).

Christie, J. R. R., 'Joseph Black and John Robison', in A. D. C. Simpson (ed.), *Joseph Black 1728–1799.*, pp. 47–52.

Clapeyron, É., 'Mémoire sur la Puissance Motrice de la Chaleur', *Journal de l'Ecole Polytechnique*, cahier 23, 14 (1834), pp. 153–190.

Collins, H., *Changing Order: Replication and Induction in Scientific Practice* (Beverly Hills, CA: SAGE Publications, 1985).

Conant, J. B., *Science and Common Sense* (London: Oxford University Press, 1951).

Corrins, R. D., 'The Great Hot-Blast Affair', *Industrial Archaeology*, 7 (1970), pp. 233–63.

Crawford, A., *Experiments and Observations on Animal Heat* (London: J. Murray, 1779, 2nd edn, 1788).

Cullen, W., 'Of the Cold Produced by Evaporating Fluids, and some other Means of Producing Cold', in *Essays and Observations. Volume II* (Edinburgh: G. Hamilton and J. Balfour, 1756).

Dalton, J., 'Experimental Essays on the Constitution of Mixed Gases; on the Force of Steam or Vapour from Water and other Liquids in Different Temperatures, Both in a Torricellian Vacuum and in Air; on Evaporation; and on the Expansion of Gases by Heat', *Memoirs and Proceedings of the Manchester Literary and Philosophical Society*, 5 (1802), pp. 535–602.

Darrigol, O., *Worlds of Flow: A History of Hydrodynamics from the Bernoullis to Prandtl* (New York: Oxford University Press, 2005).

Darwin, E., 'Frigorific Experiments on the Mechanical Expansion of Air', *Philosophical Transactions of the Royal Society of London*, 78 (1788), pp. 43–52.

—, *The Botanic Garden. A Poem in Two Parts. Part I Containing the Economy of Vegetation. Part II The Loves of the Plants. With Philosophical Notes* (London: J. Johnson, 1791).

Davies, G. L., *The Earth in Decay: A History of British Geomorphology 1578–1878* (London: Macdonald Technical and Scientific, 1969).

Davy, H., 'Experimental Essays on Heat, Light, and on the Combinations of Light', in T. Beddoes (ed.), *Contributions to Physical and Medical Knowledge, Principally from the West of England* (Bristol: Biggs & Cottle, 1799), pp. 5–150.

De Luc, J. A., *Idées sur la Météorologie*, 2 vols (London: T. Spilsbury, 1786–7).

Dean, D. R., 'Hutton Scholarship, 1992–1997', in G. Y. Craig and J. H. Hull (eds), *James Hutton - Present and Future* (London: Geological Society of London Special Publications, 1999).

Dexter, W. and J. W. T. Ley, *The Origin of Pickwick* (London: Chapman & Hall, 1936).

Dickinson, H. W., 'The Garret Workshop of James Watt', Science Museum, Technical Pamphlet No. 1 (London: HMSO, 1929).

—, *James Watt, Craftsman and Engineer* (Cambridge: Cambridge University Press, 1936).

—, *Matthew Boulton* (Cambridge: Cambridge University Press, 1936).

—, and R. Jenkins, *James Watt and the Steam Engine. The Memorial Volume Prepared for the Committee of the Watt Centenary Commemoration at Birmingham 1919* (Oxford: Clarendon Press, 1927).

Donovan, A., *Philosophical Chemistry in the Scottish Enlightenment. The Doctrines and Discoveries of William Cullen and Joseph Black* (Edinburgh: Edinburgh University Press, 1975).

—, 'James Hutton, Joseph Black and the Chemical Theory of Heat', *Ambix*, 25 (1978), pp. 176–90.

—, *Antoine Lavoisier: Science, Administration, and Revolution* (Oxford: Blackwell, 1993).

Dyer, H., 'The Steam Engine since the Days of Watt', *Papers of the Greenock Philosophical Society. – No.* 10 (Greenock: William Hutchison, 1889).

Eller, J. T. 'Seconde dissertation sur les elemens', *Mémoires de l'Academie Royal des Sciences et Belles Lettres* [Berlin], 2 (1746), pp. 25–48.

Emmerson, G. S., *John Scott Russell: A Great Victorian Naval Engineer and Naval Architect* (London: John Murray, 1977).

Engen, R. K., *Victorian Engravings*, ed. H. Best (London: St. Martins, 1975).

Ewart, P., 'On the Measure of Moving Force', *Memoirs of the Manchester Literary and Philosophical Society*, 2nd series, 2 (1813), pp. 105–258.

Ewing, J. A., 'Watt, James', *Encyclopaedia Britannica*, 9th edn, vol. 24 (1888), pp. 412–14.

Fara, P., *Newton: The Making of Genius* (Basingstoke: Macmillan, 2002).

Farrar, W. V., 'Andrew Ure, FRS, and the Philosophy of Manufactures', *Notes and Records of the Royal Society*, 27 (1972–3), pp. 299–324.

Farrar, W. V., K. R. Farrar and E. L. Scott, 'The Henrys of Manchester. Part I: Thomas Henry (1734–1816)', *Ambix*, 20 (1973), pp. 183–208.

—, 'The Henrys of Manchester. Part 2', *Ambix*, 21 (1974), pp. 179–207.

—, 'The Henrys of Manchester. Part 3', *Ambix*, 21 (1974), pp. 208–28.

—, 'The Henrys of Manchester. Part 4', *Ambix*, 22 (1975), pp. 186–204.

—, 'The Henrys of Manchester. Part 5', *Ambix*, 23 (1976), pp. 27–52.

Farson, R., *Management of the Absurd: Paradoxes in Leadership* (New York: Simon & Schuster, 1996).

Fenby, D. V., 'Chemical Reactivity and Heat in the Eighteenth Century', in P. Jones (ed.), *Philosophy and Science in the Scottish Enlightenment* (Edinburgh: John Donald Publishers, 1988), pp. 67–86.

Field, J., 'Joshua Field's Diary of a Tour in 1821 through the Midlands', Intr. J. W. Hall, *Transactions of the Newcomen Society*, 6 (1925–6), pp. 1–41.

Field, J., 'Joshua Field's Diary of a Tour in 1821 through the Provinces', Intr. E. C. Smith, *Transactions of the Newcomen Society*, 13 (1932–3), pp. 15–50.

Fleming, D., 'Latent Heat and the Invention of the Watt Engine', *Isis*, 43 (1952), pp. 3–5.

Forbes, J. D., *A Review of the Progress of Mathematical and Physical Science in More Recent Times, and Particularly between the Years 1775 and 1850; being one of the Dissertations Prefixed to the Eighth Edition of the Encyclopaedia Britannica* (Edinburgh: Adam and Charles Black, 1858).

—, *The Danger of Superficial Knowledge: An Introductory Lecture to the Course of Natural Philosophy in the University of Edinburgh, delivered on the 1st and 2nd of November, 1848* (London: John W. Parker, 1849).

Forbes, R. J., 'Power to 1850', in C. Singer et al. (eds), *A History of Technology* (Oxford: Oxford University Press, 1958).

Fox, R., *The Caloric Theory of Gases from Lavoisier to Regnault* (Oxford: The Clarendon Press, 1971).

Galison, P., *Image and Logic: A Material Culture of Microphysics* (Chicago, IL: University of Chicago Press, 1997).

Gerstner, P. A. 'James Hutton's Theory of the Earth and his Theory of Matter', *Isis*, 59 (1968), pp. 26–31.

Gieryn, T. F. *Cultural Boundaries of Science: Credibility on the Line* (Chicago, IL: University of Chicago Press, 1999).

Gillispie, C. C., *The Edge of Objectivity* (Princeton, NJ: Princeton University Press, 1960).

Goldfarb, S. J. 'Rumford's Theory of Heat: A Reassessment', *The British Journal for the History of Science*, 10 (1977), pp. 25–36.

Golinski, J., *Science as Public Culture: Chemistry and Enlightenment in Britain, 1760–1820* (Cambridge: Cambridge University Press, 1992).

—, ''Fit Instruments': Thermometers in Eighteenth-Century Chemistry', in F. L. Holmes and T. H. Levere (eds), *Instruments and Experimentation in the History of Chemistry* (Cambridge, MA: The MIT Press, 2000).

—, 'Conversations on Chemistry: Talk about Phlogiston in the Coffee House Society, 1780–87', in T. H. Levere and G. L'E. Turner (eds), *Discussing Chemistry and Steam: The Minutes of a Coffee House Philosophical Society* (Oxford: Oxford University Press, 2002), pp. 191–205.

—, *Making Natural Knowledge: Constructivism and the History of Science* (Chicago, IL: University of Chicago Press, 2005

Gooday, G. J. N., *The Morals of Measurement. Accuracy, Irony and Trust in late Victorian Electrical Practice* (Cambridge: Cambridge University Press, 2004), p. 68.

Grapí, P. and M. Izquierdo, 'Berthollet's Conception of Chemical Change in Context', *Ambix*, 44 (1997), pp. 113–30.

Gray, A. S., J. Breach, and N. Breach, *Edwardian Architecture: A Biographical Dictionary* (Iowa City, IA: University of Iowa Press, 1986).

Greenock Philosophical Society, Jubilee Celebrations (Greenock: The Philosophical Society, 1911).

Gross, A. G., 'Do Disputes over Priority Tell us Anything about Science?' *Science in Context*, 11 (1998), pp. 161–79.

Guerlac, H., 'Joseph Black's Work on Heat', in Simpson (ed.), *Joseph Black 1728 -1799*, pp. 13–22.

Hall, A. R., 'What Did the Industrial Revolution in Britain Owe to Science?', in N. McKendrick (ed.), *Historical Perspectives: Studies in English Thought and Society in Honour of J. H. Plumb* (London: Europa, 1974), pp. 129–51.

Harcourt, W. V., 'Address by the Rev. W. Vernon Harcourt', *Report of the Ninth Meeting of the British Association for the Advancement of Science held at Birmingham in August 1839* (London: John Murray, 1840), pp. 3–69.

Henry, W., *An Epitome of Chemistry, from the 4th edition* (New York: Collins and Perkins, 1808).

Herléa, A., 'Indicator', in R. Bud and D. J. Warner (eds), *Instruments of Science: An Historical Encyclopaedia* (London: Taylor & Francis, 1998), pp. 325–8.

Higgitt, R., *Recreating Newton: Newtonian Biography and the Making of Nineteenth-Century History of Science* (London: Pickering & Chatto, 2007).

Hills, R. L., *Power from Steam: A History of the Stationary Steam Engine* (1989; Cambridge: Cambridge University Press, 1993).

—, 'Houldsworth, Henry', in Lance Day and Ian McNeil (eds), *Biographical Dictionary of the History of Technology* (London: Routledge, 1996), pp. 356–7.

—, 'The Origins of James Watt's Perfect Engine', *Transactions of the Newcomen Society*, 68 (1996–7), pp. 85–107

—, 'How James Watt Invented the Separate Condenser', *Bulletin of the Scientific Instrument Society*, 57 (1998), pp. 26–9 and 58 (1998), pp. 6–10.

—, *James Watt, Volume 1: His Time in Scotland, 1736–1774* (Ashbourne: Landmark Publishing Ltd, 2002).

—, *James Watt, Volume 2: The Years of Toil, 1775–1785* (Ashbourne: Landmark Publishing Ltd, 2005).

—, *James Watt, Volume 3: Triumph through Adversity, 1785–1819* (Ashbourne: Landmark Publishing Ltd, 2006).

—, and A. J. Pacey, 'The Measurement of Power in Early Steam-driven Textile Mills', *Technology and Culture*, 13 (1972), pp. 25–43.

Hodgskin, T., 'Monument to Mr Watt', *The Chemist*, No. XVI, 26 June, 1824.

Holmes, F. L., 'Scientific Writing and Scientific Discovery', *Isis*, 78 (1987), pp. 220–5.

—, and K. M. Olesko, 'The Image of Precision: Helmholtz and the Graphical Method in Physiology', in M. N. Wise (ed.), *The Values of Precision* (Princeton, NJ: Princeton University Press, 1995), pp. 198–221.

Hopkinson, J., *The Working of the Steam Engine Explained by the Use of the Indicator* (London: Simkin, Marshall & Co., 1854).

Hutton, J., 'Theory of the Earth; or an Investigation of the Laws observable in the Composition, Dissolution, and Restoration of Land upon the Globe', *Transactions of the Royal Society of Edinburgh*, vol. 1, Pt II (1788), pp. 209–304.

—, *Dissertations on Different Subjects in Natural Philosophy* (Edinburgh: Strahan and Cadell, 1792).

—, *Theory of the Earth, with Proofs and Illustrations*, 2 vols (Edinburgh: William Creech, 1795).

Inkster, I., 'Thoughtful Doing and Early Modern Oeconomy', in L. Roberts, S. Schaffer and P. Dear (eds), *The Mindful Hand: Inquiry and Invention from the Late Renaissance to Early Industrialisation* (Amsterdam: Koninklijke Nederlandse Akademie van Wetenschappen, 2007).

Irvine, W. and W. Irvine Jr., *Essays, Chiefly on Chemical Subjects* (London: J. Mawman, 1805).

Jacob, M. C. and L. Stewart, *Practical Matter: Newton's Science in the Service of Industry and Empire, 1687–1851* (Cambridge, MA: Harvard University Press, 2004).

Jankovic, V., *Reading the Skies: A Cultural History of English Weather 1650–1820* (Manchester: Manchester University Press, 2000).

Jeffrey, F., 'The Late Mr James Watt', *The Scotsman*, 4 September 1819, p. 5.

Jeremy, D. J. *Artisans, Entrepreneurs, and Machines: Essays on the Early Anglo-American Textile Industries, 1770–1840s* (Aldershot: Ashgate, 1998).

Jones, J., H. S. Torrens and E. Robinson, 'The Correspondence between James Hutton (1726–1797) and James Watt (1736–1819) with Two Letters from Hutton to George Clerk-Maxwell (1715–1784): Part I', *Annals of Science*, 51 (1994), pp. 637–53.

Jones, P. M., 'Living the Enlightenment and the French Revolution: James Watt, Matthew Boulton and their Sons', *The Historical Journal*, 42 (1999), pp. 157–82.

—, 'Industrial Enlightenment in Practice. Visitors to the Soho Manufactory, 1765–1820', *Midland History*, 33 (2008), pp. 68–96.

—, *Industrial Enlightenment: Science, Technology and Culture in Birmingham and the West Midlands, 1760–1820* (Manchester: Manchester University Press, 2008).

Jones, R. V., 'The "Plain Story" of James Watt: The Wilkins Lecture 1969', *Notes and Records of the Royal Society of London*, 24 (1970), pp. 194–220.

Jordanova, L., *Defining Features: Scientific and Medical Portraits 1660–2000* (London: Reaktion Books/National Portrait Gallery, 2000).

Joule, J. P., 'On Some Facts in the Science of Heat Developed since the Time of Watt...Being a Lecture Delivered before the Greenock Philosophical Society on 19th January, 1865', Papers of the Greenock Philosophical Society, No. 1, printed in *Publications of the Greenock Philosophical Society* (Greenock, 1887).

Jungnickel, C. and R. McCormmach, *Cavendish. The Experimental Life* (Lewisburg, PA: Bucknell, 1999).

Kent, A., 'William Irvine, M. D. ', in idem (ed.), *An Eighteenth Century Lectureship in Chemistry* (Glasgow: Jackson, Son & company, 1950), pp. 140–50.

Keyser, B. W., 'Between Science and Craft: The Case of Berthollet and Dyeing', *Annals of Science*, 47 (1990), pp. 213–60.

Kim, M. G., 'Lavoisier, the Father of Modern Chemistry?' in M. Beretta (ed.), *Lavoisier in Perspective* (Munich: Deutsches Museum, 2005), pp. 167–91.

King, W. H. and J. W. King, *Lessons and Practical Notes on Steam, the Steam Engine ... For Young Engineers, Students and Others* (New York: Van Nostrand, 1864).

King-Hele, D. (ed.), *The Collected Letters of Erasmus Darwin* (Cambridge: Cambridge University Press, 2007).

Kirsch, D., *The Electric Vehicle and the Burden of History* (Piscataway, NJ: Rutgers University Press, 2000).

Kirwan, R., *An Essay on Phlogiston, and the Constitution of Acids. A New Edition* (London: J. Johnson, 1789).

Kosák, J., V. Moreira and D. Oldroyd, *The Iconography of the Lisbon Earthquake* (Prague: Geophysical Institute of the Academy of Sciences of the Czech Republic, 2005).

Laidler, K. J., 'James Watt and the Science of Thermodynamics', in idem, *To Light Such a Candle. Chapters in the History of Science and Technology* (Oxford: Oxford University Press, 1998).

Larder, D. F. 'An Unpublished Chemical Essay of James Watt', *Notes and Records of the Royal Society of London,* 25 (1970), pp. 193–210.

Le Van, W. B., *The Steam Engine and the Indicator: Their Origin and Progressive Development* (London: E. & F. N. Spon, 1889).

Levere, T. H., 'Dr. Thomas Beddoes and the Establishment of his Pneumatic Institution. A Tale of Three Presidents', *Notes and Records of the Royal Society of London*, 32 (1977), pp. 41–9.

—, 'Dr. Thomas Beddoes at Oxford. Radical Politics in 1788–1793 and the Fate of the Regius Chair in Chemistry', *Ambix*, 28 (1981), pp. 61–9.

—, 'Dr. Thomas Beddoes (1760–1808): Science and Medicine in Politics and Society', *The British Journal for the History of Science*, 17 (1984), pp. 187–204.

—, 'Dr Thomas Beddoes (1760–1808): Chemistry, Medicine, and Books in the French and Chemical Revolutions', in L. Principe (ed.), *New Narratives in Eighteenth-Century Chemistry* (Dordrecht: Springer, 2007), pp. 157–76.

—, and G. L'E. Turner (eds), *Discussing Chemistry and Steam. The Minutes of a Coffee House Philosophical Society 1780–1787* (Oxford: Oxford University Press, 2002).

Macaulay, T. B., *The History of England from the Accession of James the Second* (London: Longman, Brown, Green, Longmans & Roberts, 1856).

—, 'The Literature of Britain: A Speech Delivered at the Opening of the Edinburgh Philosophical Institution on the 4th of November, 1846', in *The Works of Lord Macaulay, complete, edited by Lady Trevelyan*, 8 vols (London: Longmans, Green, and Co., 1866), vol. 8, pp. 377–84.

MacLeod, C., 'Concepts of Invention and the Patent Controversy in Britain', in R. Fox (ed.), *Technological Change. Methods and Themes in the History of Technology* (Amsterdam: Harwood Academic Publishers, 1996), pp. 137–53.

—, 'James Watt, Heroic Invention and the Idea of the Industrial Revolution', in M. Berg and K. Bruland (eds), *Technological Revolutions in Europe. Historical Perspectives* (Cheltenham: Edward Elgar, 1998), pp. 96–116.

—, *Heroes of Invention: Technology, Liberalism and British Identity 1750–1914* (Cambridge: Cambridge University Press, 2007).

—, and J. Tann, 'From Engineer to Scientist: Reinventing Invention in the Watt and Faraday Centenaries, 1919–1931', *The British Journal for the History of Science,* 40 (2007), pp. 389–411.

Macquer, P. J., *Dictionnaire de Chymie* (Paris, 1766, 2nd edn, revised and enlarged, 1778).

Main, T. J. and T. Brown, *The Indicator and Dynamometer with their Practical Applications* (London: Herbert, 1847).

Marsden, B., 'Engineering Science in Glasgow: Economy, Efficiency and Measurement as Prime Movers in the Differentiation of an Academic Discipline', *The British Journal for the History of Science,* 25 (1992), pp. 319–46.

—, '"A Most Important Trespass": Lewis Gordon and the Glasgow Chair of Civil Engineering and Mechanics, 1840–55', in C. Smith and J. Agar (eds), *Making Space for Science: Territorial Themes in the Shaping of Knowledge* (Basingstoke: Macmillan Press Ltd, 1998), pp. 87–117.

—, 'Blowing Hot and Cold: Reports and Retorts on the Status of the Air-Engine as Success or Failure, 1830–1855', *History of Science*, 36 (1998), pp. 373–420.

—, *Watt's Perfect Engine: Steam and the Age of Invention* (Cambridge: Icon Books, 2002).

—, 'Engineering', in J. L. Heilbron (ed.), *The Oxford Companion to the History of Science* (New York: Oxford University Press, 2003), pp. 257–8.

—, 'Engineering Science', in J. L. Heilbron (ed.), *The Oxford Companion to the History of Science* (New York: Oxford University Press, 2003), pp. 258–61.

—, '"The Progeny of these two 'Fellows'" .... ': Robert Willis, William Whewell and the Sciences of Mechanism, Mechanics and Machinery in Early Victorian Britain', *The British Journal for the History of Science*, 37 (2004), pp. 401–34.

—, *W. J. M. Rankine, 1820–72, and the Making of Engineering Science* (Aldershot: Ashgate, forthcoming).

—, and C. Smith, *Engineering Empires: A Cultural History of Technology in Nineteenth-Century Britain* (Basingstoke: Palgrave Macmillan, 2005).

Mason, S., *The Hardware Man's Daughter: Matthew Boulton and his 'Dear Girl'* (Chichester: Phillimore, 2005).

McCormmach, R., *Speculative Truth: Henry Cavendish, Natural Philosophy, and the Rise of Modern Theoretical Science* (Oxford: Oxford University Press, 2004).

McKenzie, R., 'Hard Lessons: Public Sculpture and the Education System in Nineteenth-Century Glasgow', in G. Coutts and T. Jokela (eds), *Art, Community and Environment. Educational Perspectives* (Bristol: Intellect Books, 2008), pp. 241–62.

McKie, D. and N. H. de V. Heathcote (eds, trans.), 'William Cleghorn's *De Igne* (1779)', *Annals of Science*, 14 (1958), pp. 1–82.

Mendoza, E. (ed.), *Reflections on the Motive Power of Fire by Sadi Carnot and other Papers on the Second Law of Thermodynamics by E. Clapeyron and R. Clausius* (New York: Dover Publications, 1960), pp. 71–105.

Metzger, H., *Newton, Stahl, Boerhaave et la Doctrine Chimique* (Paris: Félix Alcan, 1930).

Middleton, W. E. K., 'Chemistry and Meteorology, 1700–1825', *Annals of Science*, 20 (1965), pp. 125–41.

—, *A History of the Theories of Rain and other Forms of Precipitation* (London: Oldbourne, 1965).

Miller, D. P., 'Between Hostile Camps: Sir Humphry Davy's Presidency of the Royal Society of London, 1820–27', *The British Journal for the History of Science*, 16 (1983), pp. 1–47.

—, 'The Usefulness of Natural Philosophy: The Royal Society of London and the Culture of Practical Utility in the later Eighteenth Century', *The British Journal for the History of Science*, 32 (1997), pp. 185–201.

—, '"Puffing Jamie": The Commercial and Ideological Importance of being a "Philosopher" in the Case of the Reputation of James Watt (1735–1819)', *History of Science*, 38 (2000), pp. 1–21.

—, *Discovering Water: James Watt, Henry Cavendish and the Nineteenth-Century 'Water Controversy'* (Aldershot: Ashgate, 2004).

—, 'True Myths: James Watt's Kettle, his Condenser, and his Chemistry', *History of Science*, 42 (2004), pp. 333–60.

—, 'Watt in Court: Specifying Steam Engines and Classifying Engineers in the Patent Trials of the 1790s', *History of Technology*, 27 (2006), pp. 43–76.

—, 'Principle, Practice and Persona in Isambard Kingdom Brunel's Patent Abolitionism', *The British Journal for the History of Science*, 41 (2008), pp. 43–72.

—, 'Seeing the Chemical Steam Through the Historical Fog: Watt's Steam Engine as Chemistry', *Annals of Science*, 65 (2008), pp. 47–72.

—, 'Scales of Justice: Assaying the Reputation of Matthew Boulton and the Partnership of Boulton and Watt', *Midland History*, 34 (2009), pp. 58–76.

— and T. H. Levere, ' "Inhale it and See?" The Collaboration between Thomas Beddoes and James Watt in Pneumatic Medicine', *Ambix*, 55 (2008), pp. 5–28.

Miller, H., *First Impressions of England and its People* (1856; Boston: Gould and Lincoln, 1860).

Morrell, J. and A. Thackray, *Gentlemen of Science: Early Years of the British Association for the Advancement of Science* (Oxford: Clarendon Press, 1981).

—, (eds), *Gentlemen of Science. Early Correspondence of the British Association for the Advancement of Science* (London: The Royal Historical Society, 1984).

Morrell, J. B., 'Professors Robison and Playfair and the *Theophobia Gallica*: Natural Philosophy, Religion and Politics in Edinburgh, 1789–1815', *Notes and Records of the Royal Society of London*, 26 (1971), pp. 43–63.

Muirhead, J. P (ed.), *Correspondence of the Late James Watt on his Discovery of the Theory of the Composition of Water* (London: John Murray, 1846).

—, *The Origin and Progress of the Mechanical Inventions of James Watt*, 3 vols (London: John Murray, 1854).

—, *The Life of James Watt with Selections from his Correspondence* (London: John Murray, 1858).

Murchison, R. I. and E. Sabine, 'Address', *Report of the Tenth Meeting of the BAAS Held at Glasgow in August 1840* (1841), pp. xxxv–xlviii on p. xxxv.

Musson, A. E. and E. Robinson, *Science and Technology in the Industrial Revolution* (Manchester: Manchester University Press, 1969).

Noszlopy, G. T. and F. Waterhouse, *Public Sculpture of Staffordshire and the Black Country* (Liverpool: Liverpool University Press, 2005).

Nye, D. E., *American Technological Sublime* (Cambridge, MA: The MIT Press, 1996).

Oldroyd, D. R., 'The Doctrine of Property Conferring Principles in Chemistry', *Organon*, 12–13 (1976–7), pp. 139–55.

—, *Thinking about the Earth: A History of Ideas in Geology* (London: Athlone, 1996).

—, 'Jean-André de Luc (1727–1817) and the Relationship between Geology and Religion: An Atheist's Comparative View of Historiography', Geological Society, London, Special Publications, 310 (2009), pp. 7–15.

Partington, J. R., *History of Chemistry*, 4 vols (London: Macmillan, 1962–4).

Patterson, W. C., *Nuclear Power*, 2nd edn (Harmondsworth: Penguin, 1983).

Peacock, G., 'Arago and Brougham on Black, Cavendish, Priestley and Watt', *Quarterly Review*, 77 (1845), pp. 105–39.

Penn, T. Z., 'The Introduction of Calico Cylinder Printing in America: A Case Study in the Transmission of Technology', in I. M. G. Quimby and P. A. Earl (eds), *Technological Innovation and the Decorative Arts* (Charlottesville, VA: University of Virginia Press, 1974), pp. 247–50.

Pickering, A., *The Mangle of Practice: Time, Agency and Science* (Chicago, IL: University of Chicago Press, 1995).

Playfair, J., *Illustrations of the Huttonian Theory of the Earth* (Edinburgh: W. Creech, 1802).

—, 'Biographical Account of the late Dr James Hutton, F. R. S. Edin. ', *Transactions of the Royal Society of Edinburgh*, 5 (1805), pp. 39–99.

Potts, A., 'Chantrey as the National Sculptor of Early Nineteenth-Century England', *Oxford Art Journal*, 4 (November 1981), pp. 17–27.

Preece, W. H., *Watt and the Measurement of Power. Being the Watt Anniversary Lecture delivered before the Greenock Philosophical Society, 5 February 1897* (London: William Clowes & Sons, 1897).

Priestley, J., 'Observations on different kinds of air', *Philosophical Transactions of the Royal Society of London*, 62 (1772), pp. 147–264.

—, *Experiments and Observations Relating to Various Branches of Natural Philosophy, with a Continuation of the Observations on Air II* (Birmingham: J. Johnson, 1781).

—, *Experiments and Observations on Different Kinds of Air...In Three Volumes*, 3rd edn (Birmingham: Thomas Pearson, 1790).

—, 'Additional Experiments and Observations Relating to the Principle of Acidity, the Decomposition of Water, and Phlogiston.... With Letters to Him on the Subject by Dr Withering and James Keir Esq. ', *Philosophical Transactions of the Royal Society of London*, 81 (1791), pp. 213–22.

—, *Experiments on the Generation of Air from Water* (London: J. Johnson, 1793).

—, *The Doctrine of Phlogiston Established, and that of the Composition of Water Refuted* (Northumberland, PA: Printed for the Author, 1800).

Prosser, R. B., *Birmingham Inventors and Inventions* (Birmingham: The "Journal" Printing Works, New Street, 1881).

Pryce, W., *Mineralogia Cornubiensis; A Treatise on Minerals, Mines and Mining* (London: J. Phillips, 1778).

Rankine, W. J. M., *Introductory Lecture on the Harmony of Theory and Practice in Mechanics, delivered to the Class of Civil Engineering and Mechanics in the University of Glasgow, on Thursday, January 3, 1856* (London and Glasgow: Richard Griffin and Company, 1856).

—, *A Manual of the Steam Engine and other Prime Movers*, 3rd edn (1859; London: Charles Griffin and Co., 1866).

Rappoport, R., 'Rouelle and Stahl–The Phlogistic Revolution in France', *Chymia*, 7 (1961), pp. 73–102.

Riskin, J., *Science in the Age of Sensibility: The Sentimental Empiricists of the French Enlightenment* (Chicago, IL: University of Chicago Press, 2002).

Roberts, L., 'An Arcadian Apparatus. The Introduction of the Steam Engine into the Dutch Landscape', *Technology and Culture*, 45 (2004), pp. 251–76.

Robinson, E., 'An English Jacobin: James Watt, Junior, 1769–1848', *Cambridge Historical Journal*, 11 (1954–5), pp. 349–55.

—, 'James Watt and the Tea Kettle: A Myth Justified', *History Today*, 6 (1956), pp. 261–5.

—, 'Matthew Boulton and the Art of Parliamentary Lobbying', *The Historical Journal*, 7 (1964), pp. 209–29.

—, 'James Watt and the Law of Patents', *Technology & Culture*, 13 (1972), pp. 115–39.

—, and A. E. Musson, *James Watt and the Steam Revolution* (London: Adams and Dart, 1969).

—, and D. McKie (eds), *Partners in Science: Letters of James Watt and Joseph Black* (Cambridge, MA: Harvard University Press, 1970).

Robison, J., 'Steam', *Encyclopaedia Britannica*, 3rd edn, 18 vols (Edinburgh: A. Bell and C. Macfarquhar, 1788–97), vol. 17, pp. 733–43.

—, 'Steam-Engine', *Encyclopaedia Britannica*, 3rd edn, 18 vols (Edinburgh: A. Bell and C. Macfarquhar, 1788–97), vol. 17, pp. 743–72.

—, *A System of Mechanical Philosophy...with Notes by David Brewster*, 4 vols (Edinburgh: John Murray, 1822).

Rudwick, M. J. S., 'Jean-André de Luc and Nature's Chronology', in C. L. E. Lewis and S. J. Knell (eds), *The Age of the Earth from 4004BC to AD2002* (London: The Geological Society, 2001), pp. 51–60.

Rumford, Count [Benjamin Thompson], 'An Inquiry Concerning the Source of the Heat which is Excited by Friction', *Philosophical Transactions of the Royal Society of London*, 88 (1798), pp. 80–102.

Russell, J. S. 'On the Application of the Inventions of Watt to Modern Steam Navigation' [The Watt Anniversary Address for 1867], Papers of the Greenock Philosophical Society – No. 3, in *Publications of the Greenock Philosophical Society* (Greenock: The Society, 1887).

Schofield, R. E., *The Lunar Society of Birmingham. A Social History of Provincial Science and Industry in Eighteenth-Century England* (Oxford: Clarendon Press, 1963).

—, (ed.), *A Scientific Autobiography of Joseph Priestley (1733–1804). Selected Scientific Correspondence Edited with Commentary* (Cambridge, MA: The M. I. T. Press, 1966.

—, *The Enlightened Joseph Priestley. A Study of His Life and Work from 1773 to 1804* (University Park, PA: Pennsylvania State University Press, 2004).

Schuster, J. A. and G. Watchirs, 'Natural Philosophy, Experiment and Discourse: Beyond the Kuhn/Bachelard Problematic', in H. E. LeGrand (ed.), *Experimental Inquiries* (Dordrecht: Kluwer, 1990), pp. 1–48.

Shapin, S., 'The Philosopher and the Chicken: On the Dietetics of Disembodied Knowledge', in S. Shapin and C. Lawrence (eds), *Science Incarnate: Historical Embodiments of Natural Knowledge* (Chicago, IL: University of Chicago Press, 1998), pp. 21–50.

Sharpe, J., 'An Account of Some Experiments to Ascertain whether the Force of Steam be in Proportion to the Generating Heat', *Memoirs of the Literary and Philosophical Society of Manchester*, Second Series, 2 (1813), pp. 1–14.

Simpson, A. D. C. (ed.), *Joseph Black 1728–1799. A Commemorative Symposium* (Edinburgh: The Royal Scottish Museum, 1982).

Smeaton, W. A., 'Some Comments on James Watt's Published Account of his Work on Steam and Steam Engines', *Notes and Records of the Royal Society of London*, 26 (1971), pp. 35–42.

Smiles, S., *Lives of Boulton and Watt* (London: John Murray, 1865).

Smith, C. ' "Nowhere but in a Great Town": William Thomson's Spiral of Classroom Credibility', in C. Smith and J. Agar (eds), *Making Space for Science: Territorial Themes in the Shaping of Knowledge* (Basingstoke: Macmillan, 1998), pp. 118–46.

—, *The Science of Energy: A Cultural History of Energy Physics in Victorian Britain* (London: The Athlone Press, 1998).

—, and M. N. Wise, *Energy and Empire. A Biographical Study of Lord Kelvin* (Cambridge: Cambridge University Press, 1989).

Smith, R. M., *The History of Greenock* (Greenock: Orr, Pollock & Co., 1921).

Sotheby's, *The James Watt Sale: Art & Science. The Property of the late Lord Gibson-Watt and Members of the Gibson-Watt Family* (London: Sotheby's, 2003).

Stansfield, D. A. and R. G. Stansfield, 'Dr Thomas Beddoes and James Watt: Preparatory Work 1794–96 for the Bristol Pneumatic Institute', *Medical History*, 30 (1986), pp. 276–302

Stock, J. E., *Memoirs of Thomas Beddoes, M. D. with an Analytical Account of his Writings* (1811; Thoemmes Press, 2003).

Suzuki, K, '*Yokohama-e* and *Kaika-e* Prints: Japanese Interpretations of Self and Other from 1860 through the 1880s', in H. Hardacre and A. L. Kern (eds), *New Directions in the Study of Meiji Japan* (Leiden: Brill, 1997), pp. 676–87.

Tait, P. G., *Heat* (London: Macmillan, 1884).

Thomsett, R., *Radical Project Management* (Upper Saddle River, NJ: Prentice Hall, 2002).

Thomson, W. (Lord Kelvin), 'James Watt: An Oration', in W. Thomson, Lord Kelvin, *Mathematical and Physical Papers*, 6 vols (Cambridge: Cambridge University Press, 1882–1911), vol 6, pp. 345–62.

Thorpe, T. E., *Essays in Historical Chemistry* (London: Macmillan & Co. Ltd, 1902).

Thurston, R. H., *A Handbook of Engine and Boiler Trials and of the Indicator and Prony Brake* (London: E. & F. N. Spon, 1890).

Torrens, H., 'Jonathan Hornblower (1753–1815) and the Steam Engine: A Historiographic Analysis', in D. Smith (ed.), *Perceptions of Great Engineers: Fact and Fantasy* (London: Science Museum for the Newcomen Society, National Museums and Galleries on Merseyside and University of Liverpool, 1994), pp. 23–34.

Tunbridge, P. A., 'Jean André De Luc, F. R. S. (1727–1817)', *Notes and Records of the Royal Society of London*, 26 (1971), pp. 15–33.

Turner, C. H. (ed.), *Proceedings of the Public Meeting held at the Freemason's Hall on the 18th June, 1824, for Erecting a Monument to ... James Watt* (London: John Murray, 1824).

Uglow, J., *The Lunar Men: The Friends Who Made the Future, 1730–1810* (London: Faber and Faber, 2002).

Ure, A., 'New Experimental Researches on Some of the Leading Doctrines of Caloric: Particularly on the Relation between the Elasticity, Temperature, and Latent Heat of Different Vapours; And on Thermometric Admeasurement and Capacity', *Philosophical Transactions of the Royal Society of London*, 108 (1818), pp. 338–94.

—, *A Dictionary of Chemistry*, 3rd edn (London: Thomas Tegg, 1828).

Vickers, N., *Coleridge and the Doctors 1795–1806* (Oxford: Clarendon Press, 2004).

Walter, J., *The Engine Indicator: A Short History of the Autographic Patterns, from 1800 to the Present Day* (Calgary: The Canadian Museum of Mining, 2008). At website http://www.archivingindustry.com/indicator/contentback.htm

Watt Centenary Committee, James Watt Centenary Commemoration, Sept. 16–20, 1919. *Souvenir Guide Book of Special Exhibit in the Art Gallery, Birmingham* (Birmingham: n. p., 1919).

Watt, J., 'Thoughts on the Constituent Parts of Water and of Dephlogisticated Air; with an Account of some Experiments on that Subject. In a Letter from Mr James Watt, Engineer, to Mr De Luc, F. R. S.', *Philosophical Transactions of the Royal Society of London*, 74 (1784), pp. 329–53.

—, 'Sequel to the Thoughts on the Constituent Parts of Water and Dephlogisticated Air. In a Subsequent Letter from Mr. James Watt, Engineer, to Mr. De Luc, F. R. S.', *Philosophical Transactions of the Royal Society of London*, 74 (1784), pp. 354–7.

Whewell, W., *The Philosophy of the Inductive Sciences*, 2 vols (London: Parker, 1840).

—, *The Mechanics of Engineering. Intended for Use in Universities, and in Colleges of Engineers* (London and Cambridge: John W. Parker and J. & J. Deighton, 1841).

[—], 'On Mr. Macaulay's Praise of Superficial Knowledge', *Fraser's Magazine*, 40 (Jul-Dec 1849), pp. 171–5.

Williamson, G., *Memorials of the Lineage, Early Life, Education and Development of the Genius of James Watt* (Edinburgh: T. Constable, 1856).

Wilson, G., *The Life of the Honourable Henry Cavendish* (London: The Cavendish Society, 1851).

Wise, M. N., 'Precision: Agent of Unity and Product of Agreement. Part II – The Age of Steam and Telegraphy', in M. N. Wise (ed.), *The Values of Precision* (Princeton, NJ: Princeton University Press, 1995), pp. 222–36.

Withering, W., 'An Analysis of Two Mineral Substances, viz. the Rowley-rag-stone and the Toad-stone', *Philosophical Transactions of the Royal Society of London*, 72 (1782), pp. 327–36.

Withers, C. W. J., *Placing the Enlightenment: Thinking Geographically about the Age of Reason* (Chicago, IL: University of Chicago Press, 2007).

Woolrich, A. P., 'John Farey and his *Treatise on the Steam Engine* (1827)', *History of Technology*, 22 (2000), pp. 63–106.

Wyke, T., *Public Sculpture in Greater Manchester* (Liverpool: Liverpool University Press, 2005).

Yeo, R., *Defining Science: William Whewell, Natural Knowledge, and Public Debate in Early Victorian Britain* (Cambridge: Cambridge University Press, 1993).

—, *Encyclopaedic Visions: Scientific Dictionaries and Enlightenment Culture* (Cambridge: Cambridge University Press, 2001).

Unpublished theses and papers

Dyck, D. R., 'The Nature of Heat and its Relationship to Chemistry in the Eighteenth Century', PhD dissertation, University of Wisconsin, 1967.

Feldmann, T. S., 'The History of Meteorology, 1750–1800: A Study in the Quantification of Experimental Physics', PhD dissertation, University of California, Berkeley, 1983.

Gibson-Watt, J., 'James Watt and the Doldowlod Estate. Extracts from the Correspondence and Notes of James Watt and James Watt Junior, 1798–1819', typescript, Archives, Birmingham Central Library.

Miller, D. P, 'Connecting Engineering and Natural Knowledge in the Royal Society', paper presented at the '3 Societies' Conference, Keble College, Oxford, July 2008.

Verbruggen, J. A., 'The Correspondence of Jan Daniël Huichelbos van Liender (1732–1809) with James Watt (1736–1819) and Boulton & Watt, Supplemented by a Few Related Documents', PhD dissertation, University of Twente, 2005.

Wise, M. N., 'Work and Waste: Political Economy and Natural Philosophy in Nineteenth-Century Britain' (unpublished essay).

INDEX

Note: Page numbers in *italics* refer to illustrations. J. W. in the index stands for James Watt.